Spatial Analysis: Theory and Practice

Spatial Analysis: Theory and Practice

Lorenzo Jade

Larsen & Keller
www.larsen-keller.com

Spatial Analysis: Theory and Practice
Lorenzo Jade
ISBN: 978-1-64172-090-8 (Hardback)

☰ Larsen & Keller

Published by Larsen and Keller Education,
5 Penn Plaza,
19th Floor,
New York, NY 10001, USA

Cataloging-in-Publication Data

Spatial analysis : theory and practice / Lorenzo Jade.
        p. cm.
Includes bibliographical references and index.
ISBN 978-1-64172-090-8
1. Geographical positions. 2. Spatial analysis (Statistics). 3. Earth sciences.
4. Geographic information systems. I. Jade, Lorenzo.
G109 .S63 2019
912--dc23

For more information regarding Larsen and Keller Education and its products, please visit the publisher's website www.larsen-keller.com

Table of Contents

Preface

Spatial analysis refers to the set of analytical approaches and techniques concerned with the study of geographical elements and entities, and their geometric, geographical and topological properties. Some of the issues in spatial analysis involve spatial characterization of entities, spatial dependency and spatial auto-correlation problem, scaling, spatial sampling, etc. The principal types of spatial analysis are spatial data analysis, spatial autocorrelation, spatial interpolation, spatial regression, and simulation and modeling, among many others. The analysis of geospatial data involves statistical analyses using software, geographic information systems (GIS) and geomatics. It has applications in diverse areas like ecology, geology, epidemiology, defense, disaster risk reduction and management, etc. This book provides comprehensive insights into the field of spatial analysis. Most of the topics introduced herein cover new techniques, practices and applications of spatial analysis. Those with an interest in this field would find this textbook helpful.

A short introduction to every chapter is written below to provide an overview of the content of the book:

Chapter 1, Spatial analysis is the technique of studying spaces or entities by analyzing their geometric, geographical or topological properties. This is an introductory chapter, which will discuss in brief all the significant aspects of spatial analysis, such as spatial association and spatial sampling; **Chapter 2**, Some of the different types of spatial analysis include spatial data analysis, spatial autocorrelation, spatial interpolation, spatial regression, spatial interaction, etc. This chapter closely examines these important spatial analysis techniques and elucidates the various processes of spatial analysis; **Chapter 3**, Geospatial analysis refers to the approach that involves the application of statistical analysis and different analytical techniques to geographical or spatial data. The aim of this chapter is to explore the fundamental concepts of geospatial analysis, such as surface analysis, network analysis and geovisualization which will closely aid in its understanding; **Chapter 4**, The science of geospatial topology is concerned with an understanding of the points, lines and polygons, which are representations of features of a geographic area. This chapter has been carefully written to provide an easy understanding of the varied facets of geospatial topology. It includes crucial topics such as spatial relation, DE-9IM, topology analysis, geography markup language, etc.; **Chapter 5**, Geostatistics is a sub-field of statistics. It studies spatial or spatiotemporal data. It has applications in diverse areas of petroleum geology, hydrology, meteorology, geochemistry, geometallurgy, etc. Some of the vital tools and techniques used in geostatistics such as variogram, kriging, geodemographic segmentation and geotargeting, etc. have been extensively discussed in this chapter; **Chapter 6**, Any database that is meant for storing and querying data representing objects in a geometric space is called a spatial database. These databases consist of geometric objects like polygons, points and lines for spatial representation. This chapter provides comprehensive knowledge of the different spatial database and database systems. It also includes an extensive discussion on object-based spatial database, spatiotemporal database and location intelligence, among others.

I extend my sincere thanks to the publisher for considering me worthy of this task. Finally, I thank my family for being a source of support and help.

Lorenzo Jade

Introduction to Spatial Analysis

Spatial analysis is the technique of studying spaces or entities by analyzing their geometric, geographical or topological properties. This is an introductory chapter, which will discuss in brief all the significant aspects of spatial analysis, such as spatial association and spatial sampling.

Spatial analysis is a type of geographical analysis which seeks to explain patterns of human behaviour and its spatial expression in terms of mathematics and geometry, that is, locational analysis.

Through spatial analysis you can interact with a GIS to answer questions, support decisions, and reveal patterns. Spatial analysis is in many ways the crux of a GIS, because it includes all of the transformations, manipulations, and methods that can be applied to geographic data to turn them into useful information.

While methods of spatial analysis can be very sophisticated, they can also be very simple. Spatial analysis is spread out along a continuum of sophistication, ranging from the simplest types that occur very quickly and intuitively when the eye and brain look at a map, to the types that require complex software and advanced mathematical knowledge.

There are many ways of defining spatial analysis, but all in one way or another express the fundamental idea that information on locations is essential. Basically, think of spatial analysis as "a set of methods whose results change when the locations of the objects being analysed change."

For example, calculating the average income for a group of people is not spatial analysis because the result doesn't depend on the locations of the people. Calculating the centre of the United States population, however, is spatial analysis because the result depends directly on the locations of residents.

Types of Spatial Analysis

- Queries and reasoning: They are the most basic of analysis operations, in which the GIS is used to answer simple questions posed by the user. No changes occur in the database and no new data are produced.

- Measurements: They are simple numerical values that describe aspects of geographic data. They include measurement of simple properties of objects, such as length, area or shape, and of the relationships between pairs of objects, such as distance or direction.

- Transformations: They are simple methods of spatial analysis that change data sets by combining them or comparing them to obtain new data sets and eventually new insights. Transformations use simple geometric, arithmetic or logical rules and they include operations that convert raster data to vector data or vice versa. They may also create fields from collections of objects or detect collections of objects in fields.

- Descriptive summaries: They attempt to capture the essence of a data set in one or two numbers. They are the spatial equivalent of the descriptive statistics commonly used in statistical analysis, including the mean and standard deviation.

- Optimization techniques: They are normative in nature, designed to select ideal locations for objects given certain well-defined criteria. They are widely used in market research, in the package delivery industry, and in a host of other applications.

- Hypothesis testing: They focus on the process of reasoning from the results of a limited sample to make generalizations about an entire population. It allows us, for example, to determine whether a pattern of points could have arisen by chance based on the information from a sample. Hypothesis testing is the basis of inferential statistics and forms the core of statistical analysis, but its use with spatial data can be problematic.

Data Types in Spatial Analysis

Consider three types of data:

- Events or point patterns: Phenomena expressed through occurrences identified as points in space, denominated point processes. Some examples are: crime spots, disease occurrences, and the localization of vegetal species.

- Continuous surfaces: Estimated from a set of field samples that can be regularly or irregularly distributed. Usually, this type of data results from natural resources survey, which includes geological, topographical, ecological, phitogeographic and pedological maps.

- Areas with Counts and Aggregated Rates: Data associated to population surveys, like census and health statistics, and that are originally referred to individuals situated in specific points in space. For confidentiality reasons these data are aggregated in analysis units, usually delimited by closed polygons (census tracts, postal addressing zones, municipalities).

Spatial Dependency

Spatial dependency is a key concept on understanding and analysing a spatial phenomena Such notion stems from what Waldo Tobler calls the first law of geography: "everything is related to everything else, but near things are more related than distant things." Or as Noel Chrissie states, "the spatial dependency is present in every direction and gets weaker the more the dispersion in the data localization increases." Generalizing we can state that most of the occurrences, natural or social, present among themselves a relationship that depends on distance. What does this principle imply? If we find pollution on a spot in a lake it is very probable that places close to this sample spot are also polluted. Or that the presence of an adult tree inhibits the development of others, such inhibition decreases with distance, and beyond a certain radius other big trees will be found.

Spatial Autocorrelation

The computational expression of the concept of spatial dependence is the spatial autocorrelation. This term comes from the statistical concept of correlation, used to measure the relationship between two random variables. The preposition "auto" indicates that the measurement of the correlation is done with the same random variable, measured in different places in space. We can use different indicators to measure the spatial autocorrelation, all of them based on the same idea: verifying how the spatial dependency varies by comparing the values of a sample and their neighbours. The autocorrelation indicators are a special case of a crossed products statistics like:

$$\Gamma(d) = \sum_{i=1}^{n}\sum_{j=1}^{n} w_{ij}(d)\xi_{ij}$$

This index expresses the relationship between different random variables as a product of two matrixes. Given a certain distance d, a matrix w_{ij} provides a measure of spatial contiguity between the random variables z_i and z_j, for example,

Informing if they are separated by a distance shorter than d. Matrix ξ_{ij} provides a measure of the correlation between these random variables that could be the product of these variables, as in the case of Moran's index for areas. And that can be expressed as:

$$I = \frac{\displaystyle\sum_{i=1}^{n}\sum_{j=1}^{n} w_{ij}(z_i - \bar{z})(z_j - \bar{z})}{\displaystyle\sum_{i=1}^{n}(z_i - \bar{z})^2}$$

Where, w_{ij} is 1 if the geographic areas associated to z_i and z_j touch each other, and 0

Otherwise, Another example of indicator is the variogram. where, we compute the square of the difference of the values, like in the case of the expression that follows:

$$\hat{\gamma}(d) = \frac{1}{2N(d)}\sum_{i=1}^{N(d)}\left[z(x_i) - z(x_i + d)\right]^2$$

Where, N(d) is the number of samples separated by distance d.

In both cases the values obtained should be compared with the values that would be produced if no spatial relationship existed between the variables. Significant values of the spatial autocorrelation indexes are evidences of spatial dependency and indicate that the postulate of independence between the samples, basis for most of the statistical inference procedures, is invalid and that the inferential models for these cases should explicitly take the space into account in its formulations.

Statistical Inference for Spatial Data

An important consequence of spatial dependence is that statistical inferences on this type of data won't be as efficient as in the case of independent samples of the same size. In other words, the spatial dependence leads to a loss of explanatory power. In general, this reflects on higher variances for the estimates, lower levels of significance in hypothesis tests and a worse adjustment for the estimated models, compared to data of the same dimension that exhibit independence.

In most cases the more adequate perspective is to consider that spatial data not as a set of independent samples, rather as one realization of a stochastic process. Contrary to the usual independent samples vision, where each observation carries independent information, in the case of a stochastic process all the observations are used in a combined way to describe the spatial pattern of the studied phenomenon. The hypothesis created in this case is that for each point u in a region A, continuous in R2, the values inferred of the attribute $z - \hat{z}(u)$– are realizations of a process $\{Z(u), u \in A\}$. In this case it is necessary to create hypothesis about the stability of the stochastic process when assuming, for example, that it is stationary and isotropic, concepts discussed in what follows.

Stationarity and Isotropy

The main statistical concepts that define the spatial structure of the data relate to the effects of 1st and 2nd order. 1st order effect is the expected value, that is, the mean of the process in space. 2nd order effect is the covariance between areas s_i and s_j. Stationarity is an important concept in this type of study. A process is considered stationary if the effects of 1st and 2nd order are constant, in the whole region under study that is there is no trend. A process is isotropic if, besides being stationary, the covariance depends only on the distance between the points and not on the direction between them. A stochastic process Z is said to be stationary of second order if the expectation of Z(u) is constant in all the region under study A, that is, it doesn't depend on its position:

$$E\{Z(u)\} = m$$

The spatial covariance structure depends solely on the relative vector between points,

$$h = u - u'$$

$$C(h) = E\{Z(u) \cdot Z(u+h)\} - E\{Z(u)\} E\{Z(u+h)\}$$

Given a specific spatial process, the stationarity hypothesis can be corroborated by explanatory analysis procedures and descriptive statistics, whose calculation should explicitly consider the spatial localization. In spatial covariance $C|h|$ the vector h comprises the distance |h| and the direction. The covariance is called anisotropic when its structure varies with distance and simultaneously as a function of its direction. When the spatial dependence is the same in all directions, we have an isotropic phenomenon.

Spatial Association

Spatial association refers to the way features are similarly distributed. When describing a spatial association, you need to identify the degree of association. A strong degree of spatial association occurs when the distribution patterns for two features are similar. A weak association describes little similarity. No association occurs when there is no similarity between the two patterns.

Spatial Sampling

Researchers who want to determine the distribution of certain properties over geographic space are usually faced with sampling limitations. For example, a mining company that wants to know the percentage content of ore in a mine can't test every inch of the mine's area to determine its contents. The company might instead use spatial sampling to test representative samples across the entirety of the mine to estimate the total value of the mine.

In spatial sampling, a number of samples are taken to determine the contents of a larger geographic area. Each sample point contains information on the variable of interest at that spatial location. The overall distribution and frequency of the variables of interest are then calculated for the entire area based on the frequency and distribution of the elements throughout the spatially sampled region.

Spatial sampling is critical for determining the contents of large areas. Studying the total contents of a large land mass is usually prohibitively expensive. Spatial sampling allows the contents instead to be inferred by studying less than 1 percent of the geographic area. Once data is collected, statisticians can use methods such as linear regression to extrapolate the overall composition of the geographic area from the information contained in the individual samples.

Potential Biases

If the contents of a study space vary at different points in the space, the area is called heterogeneous. Highly heterogeneous spaces can be difficult to study using spatial sampling; if a spatial sample misses a portion of an area that is different from the rest of the area then conclusions drawn about the whole from the sampling procedure will not be accurate. It is important to avoid sampling biases based on convenience, such as portions of an area being easier or cheaper to access.

Research Applications

Researchers can apply spatial sampling techniques to study a wide range of issues. For instance, prairie researchers use spatial sampling to determine the flora and fauna contents of entire prairies by sampling certain representative locations. These methods can also used for studying the presence of invasive or endangered species in national parks and other wildlife areas. Corporate and sociological uses for spatial sampling include determining the political views or product preference across different marketing areas.

Types of Spatial Analysis

Some of the different types of spatial analysis include spatial data analysis, spatial autocorrelation, spatial interpolation, spatial regression, spatial interaction, etc. This chapter closely examines these important spatial analysis techniques and elucidates the various processes of spatial analysis.

Spatial Data Analysis

Spatial data analysis is concerned with that branch of data analysis where the geographical referencing of objects contains important information. In many areas of data collection, especially in some areas of experimental science, the indexes that distinguish different cases can be exchanged without any loss of information. All the information relevant to understanding the variation in the data set is contained in the observations and no relevant information is contained in the indexing. In the case of spatial data the indexing (by location and time) may contain crucial information. A definition of spatial analysis (of which spatial data analysis is one element) is that it represents a collection of techniques and models that explicitly use the spatial referencing of each data case. Spatial analysis needs to make assumptions about or draw on data describing spatial relationships or spatial interactions between cases. The results of any analysis will not be the same under rearrangements of the spatial distribution of values or reconfiguration of the spatial structure.

GIS and spatial data analysis come into contact, so to speak, at the spatial data matrix. At a conceptual level, this matrix consists of rows and columns where rows refer to cases and columns refer to the attributes measured at each of the cases, and the last columns provide the spatial referencing. At the simplest level, there might be two last columns containing a pair of coordinates: latitude and longitude, or x and y in some projected coordinate system. But today database technology allows a single conceptual column to contain a complex representation of the case's spatial geometry or shape.

This conceptual matrix is but a slice through a larger cube where the other axis is time. At a practical level, the spatial data matrix is the repository of the data collected by the researcher. In practical terms the structure and content of the matrix is the end product of processes of conceptualisation and representation by which some segment of geographical reality is captured. It is in one sense the output of a process of digitally capturing the world. In another sense, the matrix is the starting point or input for the spatial data analyst.

Those principally concerned with data analysis need to give careful consideration to how well the data matrix captures the geographic reality underlying the problem and the implications (for interpreting findings) of representational choices. The question of what is missing from the representation – the uncertainty that the representation leaves in the mind of its users about the world being represented – may be as important as the content in some applications. By the same token those principally concerned with the digital representation of geographical spaces need to be aware of the power of statistical methodology to reveal useful data insights and understandings if data are made available in appropriate forms and subject to appropriate methods of analysis.

Geostatistical Tools for Spatial Analysis

Geostatistics studies spatial variability of regionalized variables: Variables that have an attribute value and a location in a two or three-dimensional space. Tools to characterize the spatial variability are:

- Spatial Autocorrelation Function.
- Variogram.

A variogram is calculated from the variance of pairs of points at different separation. For several distance classes or lags, all point pairs are identified which matches that separation and the variance is calculated. Repeating this process for various distance classes yields a variogram. These functions can be used to measure spatial variability of point data but also of maps or images.

Spatial Auto-correlation of Point Data

The statistical analysis referred to as spatial auto-correlation, examines the correlation of a random process with itself in space. Many variables that have discrete values measured at several specific geographic positions (i.e., individual observations can be approximated by dimensionless points) can be considered random processes and can thus be analysed using spatial auto-correlation analysis. Examples of such phenomena are: total amount of rainfall, toxic element concentration, grain size, elevation at triangulated points, etc.

The spatial auto-correlation function is referred to as spatial auto-correlogram, showing the correlation between a series of points or a map and itself for different shifts in space or time. It visualizes the spatial variability of the phenomena under study. In general, large numbers of pairs of points that are close to each other on average have a lower variance (i.e., are better correlated), than pairs of points at larger separation. The auto-correlogram quantifies this relationship and allows gaining insight into the spatial behaviour of the phenomenon under study.

Point Interpolation

A point interpolation performs an interpolation on randomly distributed point values and returns regularly distributed point values. The various interpolation methods are: Voronoi Tesselation, moving average, trend surface and moving surface.

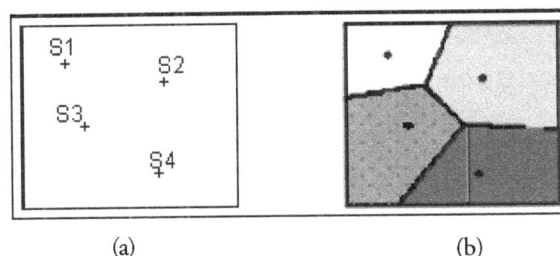

(a)   (b)

Figure: (a) An input point map, (b) The output map obtained as the result of the interpolation operation applying the Voronoi Tessellation method

Example: Nearest Neighbor (Voronoi Tessellation)-In this method the value, identifier, or class name of the nearest point is assigned to the pixels. It offers a quick way to obtain a Thiessen map from point data.

Vector Based Spatial Data Analysis

There are multilayer operations, which allow combining features from different layers to form a new map and give new information and features that were not present in the individual maps.

Topological overlays: Selective overlay of polygons, lines and points enables the users to generate a map containing features and attributes of interest, extracted from different themes or layers. Overlay operations can be performed on both and vector maps. In case of raster map calculation tool is used to perform overlay. In topological overlays polygon features of one layer can be combined with point, line and polygon features of a layer.

Polygon-in-polygon overlay:

- Output is polygon coverage.
- Coverages are overlaid two at a time.
- There is no limit on the number of coverages to be combined.
- New File Attribute Table is created having information about each newly created feature.

Line-in-polygon overlay:

- Output is line coverage with additional attribute.
- No polygon boundaries are copied.
- New arc-node topology is created.

Point-in-polygon overlay:

- Output is point coverage with additional attributes.
- No new point features are created.
- No polygon boundaries are copied.

Logical Operators

Overlay analysis manipulates spatial data organized in different layers to create combined spatial features according to logical conditions specified in Boolean algebra with the help of logical and conditional operators. The logical conditions are specified with operands (data elements) and operators (relationships among data elements).

Note: In vector overlay, arithmetic operations are performed with the help of logical operators. There is no direct way to it.

Common logical operators include AND, OR, XOR (Exclusive OR), and NOT. Each operation is characterized by specific logical checks of decision criteria to determine if a condition is true or false. Table below shows the true/false conditions of the most common Boolean operations. In this table, A and B are two operands. One (1) implies a true condition and zero (0) implies false. Thus, if the A condition is true while the B condition is false, then the combined condition of A and B is false, whereas the combined condition of A OR B is true.

AND   -   Common Area/ Intersection / Clipping Operation

OR   -   Union or Addition

NOT   -   (Inverter)

XOR   -   Minus

Table: Truth Table of common Boolean operations.

A	B	A AND B	A OR B	A NOT B	B NOT A	A XOR B
0	0	0	0	0	0	0
0	1	0	1	0	1	1
1	0	0	1	1	0	1
1	1	1	1	0	0	0

The most common basic multilayer operations are union, intersection and identify operations. All three operations merge spatial features on separate data layers to create new features from the original coverage. The main difference among these operations is in the way spatial features are selected for processing.

Overlay Operations

The figure shows different types of vector overlay operations and gives flexibility for geographic data manipulation and analysis. In polygon overlay, features from two map coverages are geometrically intersected to produce a new set of information, attributes for these new features are derived from the attributes of both the original coverages, thereby contain new spatial and attribute data relationships.

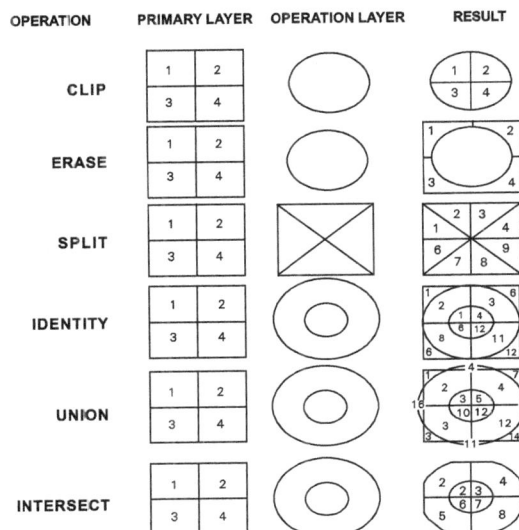

Figure: Overlay operations

One of the overlay operation is AND (or INTERSECT) in vector layer operations, in which two coverages are combined. Only those features in the area common to both are preserved. Feature attributes from both coverages are joined in the output coverage.

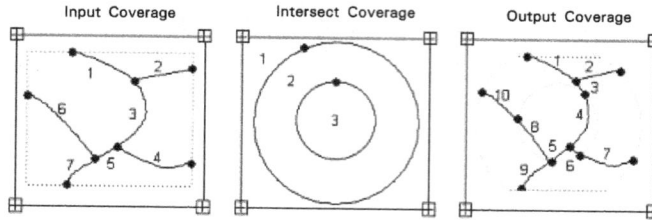

INPUT COVERAGE	
#	ATRIBUTE
1	A
2	B
3	A
4	C
5	A
6	D
7	A

INTERSECT COVERAGE	
#	ATRIBUTE
1	
2	102
3	103

OUTPUT COVERAGE	INPUT COVERAGE		INTERSECT COVERAGE	
#	#	ATRIBUTE	#	ATRIBUTE
1	1	A	2	102
2	2	B	2	102
3	3	A	2	102
4	3	A	3	103
5	5	A	3	103
6	4	C	3	103
7	4	C	2	102
8	6	D	3	103
9	7	A	2	102
10	6	D	2	102

Raster based Spatial Data Analysis

Here, we discuss operational procedures and quantitative methods for the analysis of spatial data in raster format. In raster analysis, geographic units are regularly spaced, and the location of each unit is referenced by row and column positions. Because geographic units are of equal size and identical shape, area adjustment of geographic units is unnecessary and spatial properties of geographic entities are relatively easy to trace. All cells in a grid have a positive position reference, following the left-to-right and top-to-bottom data scan. Every cell in a grid is an individual unit and must be assigned a value. Depending on the nature of the grid, the value assigned to a cell can be an integer or a floating point. When data values are not available for particular cells, they are described as NODATA cells. NODATA cells differ from cells containing zero in the sense that zero value is considered to be data.

The regularity in the arrangement of geographic units allows for the underlying spatial relationships to be efficiently formulated. For instance, the distance between orthogonal neighbours (neighbours on the same row or column) is always a constant whereas the distance between two diagonal units can also be computed as a function of that constant. Therefore the distance between any pair of units can be computed from differences in row and column positions. Furthermore, directional information is readily available for any pair of origin and destination cells as long as their positions in the grid are known.

Advantages of Using the Raster Format in Spatial Analysis

- Efficient processing: Because geographic units are regularly spaced with identical spatial properties, multiple layer operations can be processed very efficiently.

- Numerous existing sources: Grids are the common format for numerous sources of spatial information including satellite imagery, scanned aerial photos, and digital elevation models, among others. These data sources have been adopted in many GIS projects and have become the most common sources of major geographic databases.

- Different feature types organized in the same layer: For instance, the same grid may consist of point features, line features, and area features, as long as different features are assigned different values.

Grid Format Disadvantages

- Data redundancy: When data elements are organized in a regularly spaced system, there is a data point at the location of every grid cell, regardless of whether the data element is needed or not. Although, several compression techniques are available, the advantages of gridded data are lost whenever the gridded data format is altered through compression. In most cases, the compressed data cannot be directly processed for analysis. Instead, the compressed raster data must first be decompressed in order to take advantage of spatial regularity.

- Resolution confusion: Gridded data give an unnatural look and unrealistic presentation unless the resolution is sufficiently high. Conversely, spatial resolution dictates spatial properties. For instance, some spatial statistics derived from a distribution may be different, if spatial resolution varies, which is the result of the well-known scale problem.

- Cell value assignment difficulties: Different methods of cell value assignment may result in quite different spatial patterns.

Grid Operations used in Map Algebra

Common operations in grid analysis consist of the following functions, which are used in Map Algebra to manipulate grid files. The Map Algebra language is a programming language developed to perform cartographic modeling. Map Algebra performs following four basic operations:

- Local functions: that work on every single cell,

- Focal functions: that process the data of each cell based on the information of a specified neighborhood,

- Zonal functions: that provide operations that work on each group of cells of identical values,

- Global functions: that work on a cell based on the data of the entire grid.

The principal functionality of these operations:

Local Functions

Local functions process a grid on a cell-by-cell basis, that is, each cell is processed based solely on its own values, without reference to the values of other cells. In other words, the output value is a function of the value or values of the cell being processed, regardless of the values of surrounding cells. For single layer operations, a typical example is changing the value of each cell by adding or multiplying a constant. In the following example, the input grid contains values ranging from 0 to 4. Blank cells represent NODATA cells. A simple local function multiplies every cell by a constant of 3. The results are shown in the output grid at the right. When there is no data for a cell, the corresponding cell of the output grid remains a blank.

Figure: A local function multiplies each cell in the input grid by 3 to produce the output grid

Local functions can also be applied to multiple layers represented by multiple grids of the same geographic area.

Figure: A local function multiplies the input grid by the multiplier grid to produce the output grid.

Local functions are not limited to arithmetic computations. Trigonometric, exponential, and logarithmic and logical expressions are all acceptable for defining local functions.

Focal Functions

Focal functions process cell data depending on the values of neighbouring cells. For instance, computing the sum of a specified neighbourhood and assigning the sum to the corresponding cell of the output grid is the "focal sum" function. A 3 x 3 kernel defines neighbourhood. For cells closer to the edge where the regular kernel is not available, a reduced kernel is used and the sum is computed accordingly. For instance, a 2 x 2 kernel adjusts the upper left corner cell. Thus, the sum of the four values, 2,0,2 and 3 yields 7, which becomes the value of this cell in the output grid. The value of the second row, second column, is the sum of nine elements, 2, 0, 1, 2, 3, 0, 4, 2 and 2, and the sum equals 16.

Input Grid					Output Grid			
2	0	1	1		7	8	9	6
2	3	0	4	Focal Sum =	13	16	16	11
4	2	2	3		13	18	20	14
1	1	3	2		8	13	13	10

Figure: A Focal sum function sums the values of
the specified neighbourhood to produce output grid

Another focal function is the mean of the specified neighbourhood, the "focal mean" function. In the following example, this function yields the mean of the eight adjacent cells and the centre cell itself. This is the smoothing function to obtain the moving average in such a way that the value of each cell is changed into the average of the specified neighbourhood.

Input Grid					Output Grid			
2	0	1	1		1.8	1.3	1.5	1.5
2	3	0	4	Focal Mean =	2.2	2.0	1.8	1.8
4	2	2	3		2.2	2.0	2.2	2.3
1	1	3	2		2.0	2.2	2.2	2.5

Figure: A Focal mean function computes the moving average
of the specified neighbourhood to produce the output grid

Other commonly employed focal functions include standard deviation (focal standard deviation), maximum (focal maximum), minimum (focal minimum), and range (focal range).

Zonal Functions

Zonal functions process the data of a grid in such a way that cell of the same zone are analysed as a group. A zone consists of a number of cells that may or may not be contiguous. A typical zonal function requires two grids – a zone grid, which defines the size, shape and location of each zone, and a value grid, which is to be processed for analysis. In the zone grid, cells of the same zone are coded with the same value, while zones are assigned different zone values.

Figure illustrates an example of the zonal function. The objective of this function is to identify the zonal maximum for each zone. In the input zone grid, there are only three zones with values ranging from 1 to 3. The zone with a value of 1 has five cells, three at the upper right corner and two at the lower left corner. The procedure involves finding the maximum value among these cells from the value grid.

	Zone Grid				Value Grid					Output Grid			
	2	2	1	1	1	2	3	4		5	5	8	8
Zonal	2	3	3	1	5	6	7	8		5	7	7	8
Max [		3	2		1	2	3	4	]=		7	5	
	1	1	2	2	5	5	5	5		8	8	5	5

Figure: A Zonal maximum function identifies the maximum of each zone to produce the output grid

Typical zonal functions include zonal mean, zonal standard deviation, zonal sum, zonal minimum, zonal maximum, zonal range, and zonal variety. Other statistical and geometric properties may also be derived from additional zonal functions. For instance, the zonal perimeter function calculates the perimeter of each zone and assigns the returned value to each cell of the zone in the output grid.

Global Functions

For global functions, the output value of each cell is a function of the entire grid. As an example, the Euclidean distance function computes the distance from each cell to the nearest source cell, where source cells are defined in an input grid. In a square grid, the distance between two orthogonal neighbours is equal to the size of a cell, or the distance between the centroid locations of adjacent cells. Likewise, the distance between two diagonal neighbours is equal to the cell size multiplied by the square root of 2. Distance between non-adjacent cells can be computed according to their row and column addresses.

In figure, the grid at the left is the source grid in which two clusters of source cells exist. The source cells labelled 1 are the first clusters, and the cell labelled 2 is a single-cell source. The Euclidean distance from any source cell is always equal to 0. For any other cell, the output value is the distance from its nearest source cell.

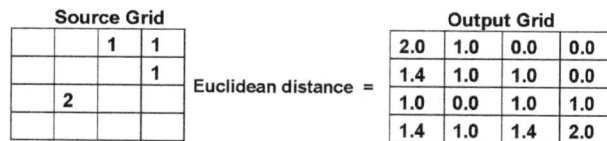

Source Grid					Output Grid			
		1	1		2.0	1.0	0.0	0.0
			1	Euclidean distance =	1.4	1.0	1.0	0.0
	2				1.0	0.0	1.0	1.0
					1.4	1.0	1.4	2.0

Figure: A Euclidean distance function computes
the distance from the nearest source cell

In the above example, the measurement of the distance from any cell must include the entire source grid; therefore this analytical procedure is a global function.

Figure below an example of the cost distance function. The source grid is identical to that in the preceding illustration. However, this time a cost grid is employed to weigh travel cost. The value in each cell of the cost grid indicates the cost for traveling through that cell. Thus, the cost for traveling from the cell located in the first row, second column to its adjacent source cell to the right is half the cost of traveling through itself plus half the cost of traveling through the neighbouring cell.

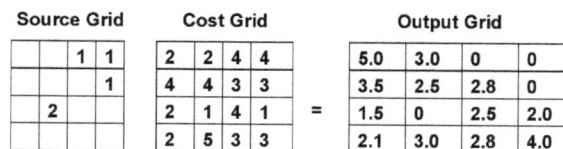

Source Grid				Cost Grid					Output Grid			
		1	1	2	2	4	4		5.0	3.0	0	0
			1	4	4	3	3		3.5	2.5	2.8	0
	2			2	1	4	1	=	1.5	0	2.5	2.0
				2	5	3	3		2.1	3.0	2.8	4.0

Figure: Travel cost for each cell is derived from the distance
to the nearest source cell weighted by a cost function

Another useful global function is the cost path function, which identifies the least cost path from each selected cell to its nearest source cell in terms of cost distance. These global functions are particularly useful for evaluating the connectivity of a landscape and the proximity of a cell to any given entities.

Some Important Raster Analysis Operations

Some of the important raster based analysis:

- Renumbering Areas in a Grid File.

- Performing a Cost Surface Analysis.

- Performing an Optimal Path Analysis.

- Performing a Proximity Search.

Area Numbering: Area Numbering assigns a unique attribute value to each area in a specified grid file. An area consists of two or more adjacent cells that have the same cell value or a single cell with no adjacent cell of the same value. To consider a group of cells with the same values beside each other, a cell must have a cell of the same value on at least one side of it horizontally or vertically (4-connectivity), or on at least one side horizontally, vertically, or diagonally (8-connectivity). Figure shows a simple example of area numbering.

Input grid   **Resultant 4 connected**

1	0	0	0	0	1		1	0	0	0	0	2
0	1	1	0	1	1		0	3	3	0	2	2
		0	1	1			4	0	3	0	2	2
		0	0	0			0	0	0	0	0	0
		0	1	1			5	5	5	0	6	6

Figure: Illustrates simple example of Area numbering with a bit map as input. The pixels, which are connected, are assigned the same code.

One can renumber all of the areas in a grid, or you can renumber only those areas that have one or more specific values. If you renumber all of the areas, Area Number assigns a value of 1 to the first area located. It then assigns a value of 2 to the second area, and continues this reassignment method until all of the areas are renumbered. When you renumber areas that contain a specified value (such as 13), the first such area is assigned the maximum grid value plus 1. For example, if the maximum grid value is 25, Area Number assigns a value of 26 to the first area, a value of 27 to the second area, and continues until all of the areas that contain the specified values are renumbered.

Cost Surface Analysis: Cost Surface generates a grid in which each grid cell represents the cost to travel to that grid cell from the nearest of one or more start locations. The cost of traveling to a given cell is determined from a weight grid file. Zero Weights option uses attribute values of 0 as the start locations. The By Row/Column option uses the specified row and column location as the start location.

Optimal Path: Optimal Path lets us analyse a grid file to find the best path between a specified location and the closest start location as used in generating a cost surface. The computation is based on a cost surface file that you generate with cost Surface.

One must specify the start location by row and column. The zeros in the input cost surface represent one endpoint. The specified start location represents the other endpoint.

Testing the values of neighbouring cells for the smallest value generates the path. When the smallest value is found, the path moves to that location, where it repeats the process to move the next cell. The output is the path of least resistance between two points, with the least expensive, but not necessarily the straightest, line between two endpoints. The output file consists of only the output path attribute value, which can be optionally specified, surrounded by void values.

Performing, A Proximity Search: Proximity lets you search a grid file for all the occurrences of a cell value or a feature within either a specified distance or a specified number of cells from the origin.

You can set both the origin and the target to a single value or a set of values. The number of cells to find can also be limited. For example, if you specify to find 10 cells, the search stops when 10 occurrences of the cell have been found within the specified distance of each origin value. If you do not limit the number of cells, the search continues until all target values are located.

The output grid file has the user-type code and the data-type code of the input file. The gird-cell values in the output file indicate whether the grid cell corresponds to an origin value, the value searched for and located within the specified target, or neither of these.

The origin and target values may be retained as the original values or specified to be another value.

Grid based Spatial Analysis

- Diffusion modelling and Connectivity analysis, can be effectively conducted from grid data. Grid analysis is suitable for these types of problems because of the grid's regular spatial configuration of geographic units.

- Diffusion Modelling: It deals with the process underlying spatial distribution. The constant distance between adjacent units makes it possible to simulate the progression over geographic units at a consistent rate. Diffusion modelling has a variety of possible applications, including wildfire management, disease vector tracking, migration studies, and innovation diffusion research, among others.

- Connectivity Analysis: Connectivity analysis evaluates inter-separation distance, which is difficult to calculate in polygon coverage, but can be obtained much more effectively in a grid.

The connectivity of a landscape measures the degree to which surface features of a certain type are connected. Landscape connectivity is an important concern in environmental management. In some cases, effective management of natural resources requires maximum connectivity of specific features. For instance, a sufficiently large area of dense forests must be well connected to provide a habitat for some endangered species to survive. In such cases, forest management policies must be set to maintain the highest possible level to connectivity. Connectivity analysis is especially useful for natural resource and environmental management.

Spatial Autocorrelation

Spatial autocorrelation in GIS helps understand the degree to which one object is similar to other nearby objects. Moran's I (Index) measures spatial autocorrelation.

Geographer Waldo R. Tobler's stated in the first law of geography:

"Everything is related to everything else, but near things are more related than distant things."

Spatial autocorrelation definition measures how much close objects are in comparison with other close objects. Moran's I can be classified as positive, negative and no spatial auto-correlation:

- Positive spatial autocorrelation is when similar values cluster together in a map.

- Negative spatial autocorrelation is when dissimilar values cluster together in a map.

Importance of Spatial Autocorrelation

One of the main reasons why spatial auto-correlation is important is because statistics relies on observations being independent from one another. If autocorrelation exists in a map, then this violates the fact that observations are independent from one another.

Another potential application is analysing clusters and dispersion of ecology and disease.

Is the disease an isolated case or spreading with dispersion?

These trends can be better understood using spatial autocorrelation analysis.

Positive Spatial Autocorrelation Example

Positive spatial auto-correlation occurs when Moran's I is close to +1. This means values cluster together. For example, elevation datasets have similar elevation values close to each other.

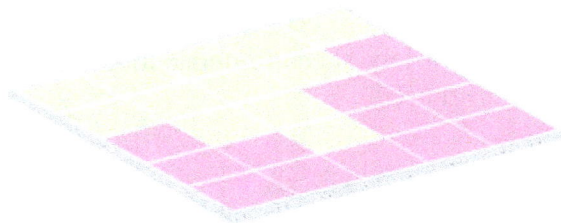

Clustered Image Spatial Autocorrelation

There is clustering in the land cover image above. This clustered pattern generates a Moran's I of 0.60. The z-score of 4.95 indicates there is less than 1% likelihood that this clustered pattern could be the result of random choice.

Negative Spatial Autocorrelation Example

Negative spatial auto-correlation occurs when Moran's I is near -1. A checkerboard is an example where Moran's I is -1 because dissimilar values are next to each other. A value of 0 for Moran's I typically indicates no autocorrelation.

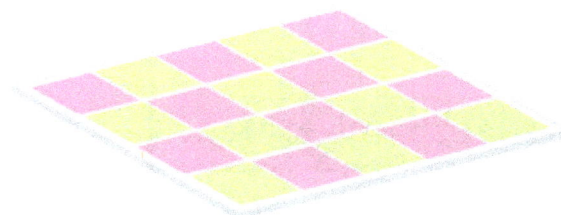

Check board Pattern: Spatial Autocorrelation

Using the spatial autocorrelation tool in ArcGIS, the checkerboard pattern generates a Moran's index of -1.00 with a z-score of -7.59.

(Remember that the z-score indicates the statistical significance given the number of features in the dataset).

Types of Common Spatial Autocorrelation in GIS Software

Perhaps the most common way in which autocorrelation is measured is using Moran's I, which now has become incorporated in commonly used packages such as ArcGIS as well as open source software such as GRASS and QGIS. Moran's I allows the correlation measure to measure how well something correlates based on multiple dimensions across a given space. Results are generally used to measure how well an object correlates globally, that is across a given defined space for a dataset. Geary's ratio (or C) is another similar measure, where this measure is more sensitive to local variations and can be used to define local patterning within a dataset.

Spatial Autocorrelation is Used to Understand Associations

Studies applying spatial autocorrelation have shown strong association with factors such as language and species diversity. Land use and land cover types show strong correlation results. Autocorrelation has also been utilized to look at the effects of health care and survival rates based on spatial-based factors. More recently, economists, who have been relatively late in utilizing spatial regression and autocorrelation techniques in econometric measures, have now also utilized spatial autocorrelation to investigate a variety of econometric indicators, including where traditional regression analyses would have been used.

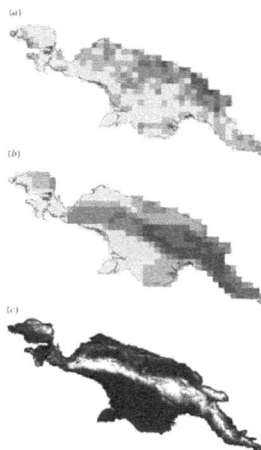

Figure: Spatial distributions of (a) language richness, (b) mammal richness and (c) elevation across the New Guinea mainland.

Spatial Stratified Heterogeneity

Spatial heterogeneity refers to the uneven distribution of a trait, event, or relationship across a region. It is frequently introduced simultaneously with the term spatial dependence and in practice the two can be difficult to tease apart from each other. Whether we are dealing with spatial heterogeneity or spatial dependence can be influenced by the scale that we are considering and the research questions in which we are interested. Spatial heterogeneity is also sometimes referred to as "sub-regional variation," "parent contagion," "first-order variation," or "first-order spatial effects." Conversely, spatial dependence is sometimes referred to as "second-order spatial effects."

Spatial heterogeneity generally refers to the clumpy or patchy distribution of processes or events across a broad landscape, whereas spatial dependence refers to processes that create clusters of events, etc. That is, spatial heterogeneity describes a patchy landscape and spatial dependence refers to the local non-independence of occurrences that are near each other.

One way to think about these two concepts is to ask yourself:

a) Is the intensity of occurrence of an event equally distributed across the landscape?

b) Does the intensity at one location influence the intensity at neighbouring locations?

If you answered yes to the first question you are dealing with spatial heterogeneity and if you answered yes to the second question you are dealing with spatial dependency. Furthermore, if a landscape has spatial heterogeneity it may be the result of spatial dependency. When considering demographic processes, spatial heterogeneity, rather than spatial homogeneity, is generally the norm. Even homogeneous environments are likely to be heterogeneous if we consider a different, larger scale (also consider the modifiable areal unit problem).

For example, within the populations that are spread across landscapes there exist pockets of low and high fertility, mortality, and population movement or migration. Sometimes events such as births or deaths are clustered. While there is variation within all sub-regions, like things do tend to coexist in similar environments. Demographers are frequently interested in uncovering, describing, and explaining such interesting clusters. There are a variety of exploratory tools for detecting spatial heterogeneity. Moran's I is often used as an indicator of spatial association.

As a very simple visual example, figure shows estimated population density (person per square kilo meter) in Thailand. It should not be surprising that there are clusters of high density in some places and areas that have very low density in between. As previously mentioned, spatial heterogeneity is the norm. In this case, we would not expect that people will be distributed randomly and homogeneously across the landscape.

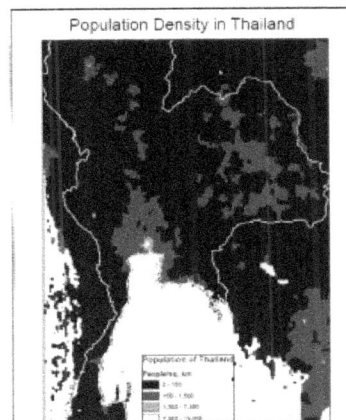
Population Density in Thailand

Further, when demographers attempt to investigate the occurrence of certain population phenomena, they frequently are interested in factors that contribute to the occurrence of these clusters. For example, within a single state there can be large disparities in fertility rates. Perhaps fertility rates are on average lower in urban areas than rural areas. Additionally, cultural norms may influence fertility rates, meaning that even within urban areas there can be subgroups of the population that both live near each other (ethnic enclaves) and have higher fertility than their immediate urban neighbours. In such a case, one way to deal with spatial heterogeneity in an explanatory model is to include variables that explain this heterogeneity—such as rural/urban status and cultural groups.

However, spatial heterogeneity does not just describe the distribution of events of interest across space; it can also refer to the distribution of a relationship across a landscape. In essence, what matters in social processes may not be homogeneously distributed across a landscape. Drawing from the previous example of fertility, it is possible that predictors of high fertility vary across space. For example, cultural group may have no predictive power with regard to high fertility rates in a rural setting, although it may have a high predictive power in an urban setting. In some cases, the effect of a predictor on its outcome can even change direction (e.g., negative to positive or vice versa) depending on location. A relatively recent approach to addressing this problem (and potentially spatial dependence as well) is geographically weighted regression.

Methods of Determining Stratified Heterogeneity in GIS

Within this approach, various methods are used to determine heterogeneity. One method is the application of spatial beta binomial distribution, where randomness is assessed as part of a Bayesian approach to seeing if given differences in a strata are evident that differ from expected probability. Other methods include using a q-statistic method that measures heterogeneity in layers in a ratio from 0 to 1, where 1 represents perfect heterogeneity and 0 is no significant heterogeneity. Techniques also combine the so-called sandwich method when null or weak values within layers might be present. In this case, stratification is created based on variance within a layer that is minimized while variation between layers is maximized. Common values are then estimated for each layer. The stratification uses units that report variation where measurements are then made to show how heterogeneous layers are based on the minimizing and maximizing approach within and between layers respectively. In effect, this is a way to classify and then measure given data to address having null or weak values present within datasets.

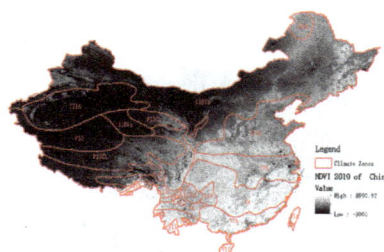

Spatial stratification of the NDVI in climate zones in Mainland China.

Applications of Stratified Heterogeneity in GIS

Examples of applications have included classifying land use systems around the globe. A common problem is that spatial variation is evident at different scales, creating problems of sampling in order to determine variety within and between strata. Additionally, adding time-dependent data can make it difficult to see how variation occurs based on time since often land use relationships are evident in one instant but not in other temporal resolution due to data collection and other issues. One way around this problem is to applying sampling techniques of land cover at a given spatial and temporal scales and then using interpolative modelling so that the remainder of a given land cover distribution can be determined from available samples at a given time. Marine environments and water quality can also be monitored using stratified heterogeneity methods, where water quality affected by phosphates and pollutants is likely to have differential effects on marine and other aquatic-related measures for water quality. While most methods apply spatial heterogeneity to landscape or biological-based examples, some have attempted to apply the approach to the social sciences. For instance, to understand disability employment in China, different regions were assessed for their qualities, including employment and population. The analysis showed biases in employment of disabled people, where such results could be used to guide policy makers in determining where resources should be spent or laws more strictly applied.

Overall, stratified heterogeneity is not a widely used method as of yet. However, it has shown potential and utility in determining patchiness within and between layers using common statistical techniques. This provides analysis with a useful way to determine the degree of spatially variation within datasets.

Spatial Interpolation

Spatial interpolation is the process of using points with known values to estimate values at other unknown points. For example, to make a precipitation (rainfall) map for your country, you will not find enough evenly spread weather stations to cover the entire region. Spatial interpolation can estimate the temperatures at locations without recorded data by using known temperature readings at nearby weather stations. This type of interpolated surface is often called a statistical surface. Elevation data, precipitation, snow accumulation, water table and population density are other types of data that can be computed using interpolation.

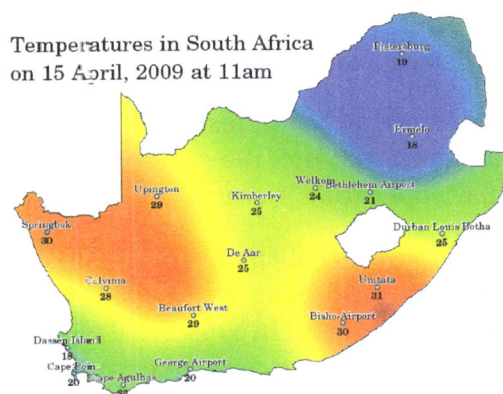

Temperature map interpolated from South African Weather Stations.

Because of high cost and limited resources, data collection is usually conducted only in a limited number of selected point locations. In GIS, spatial interpolation of these points can be applied to create a raster surface with estimates made for all raster cells.

In order to generate a continuous map, for example, a digital elevation map from elevation points measured with a GPS device, a suitable interpolation method has to be used to optimally estimate the values at those locations where no samples or measurements were taken. The results of the interpolation analysis can then be used for analyses that cover the whole area and for modelling.

There are many interpolation methods. In this introduction we will present two widely used interpolation methods called Inverse Distance Weighting (IDW) and Triangulated Irregular Networks (TIN).

Problem Formulation and Criteria for Solutions

The general formulation of the spatial interpolation problem can be defined as follows:

Given the N values of a studied phenomenon $z_j, j = 1, \dots , N$ measured at discrete points $r_j = (x_j , x_j , \dots , x_j), j = 1, \dots , N$ within a certain region of a d-dimensional space, find a d-variate function F(r) which passes through the given points, that means, fulfils the condition:

$$F(r_j) = z_j, j = 1, \dots ,N$$

Because there exist an infinite number of functions which fulfil this requirement, additional conditions have to be imposed, defining the character of various interpolation techniques. Typical examples are conditions based on geo statistical concepts (Kinging), locality (nearest neighbour and finite element methods), smoothness and tension (splines), or ad hoc functional forms (polynomials, multi-quadrics). Choice of the additional condition depends on the character of the modelled phenomenon and the type of application Finding appropriate interpolation methods for GIS applications poses several challenges. The modelled fields are usually very complex, data are spatially heterogeneous and often based on far from optimal sampling, and significant noise or discontinuities can be present. In addition, datasets can be very large (N ≈ 103–106), originating from various sources with different accuracies. Reliable interpolation tools, suitable for GIS applications, should therefore satisfy several important demands: accuracy and predictive power, robustness and flexibility in describing various types of phenomena, smoothing for noisy data, d dimensional formulation, direct estimation of derivatives (gradients, curvatures), applicability to large datasets, computational efficiency, and ease of use.

Currently, it is difficult to find a method which fulfils all of the above-mentioned requirements for a wide range of geo referenced data. Therefore, the selection of an adequate method with appropriate parameters for a particular application is crucial.

Different methods can produce quite different spatial representations and in depth knowledge of the phenomenon is needed to evaluate which one is the closest to reality. The use of an unsuitable method or inappropriate parameters can result in a distorted model of spatial distribution, leading to potentially wrong decisions based on misleading spatial information. An inappropriate interpolation can have even more profound impact if the result is used as an input for simulations, where a small error or distortion can cause models to produce false spatial patterns, for a discussion of error propagation. Quantitative evaluation of interpolation predictive capabilities, for example by cross-validation, is often not sufficient for the selection of an appropriate interpolation method, as the preservation of geometrical properties is in some cases more important than actual accuracy.

Advanced visualisation and analysis of slope, aspect, and curvature is helpful in detecting geometrical distortions.

Methods

In recent years, GIS capabilities for spatial interpolation have improved by integration of advanced methods within GIS, as well as by linking GIS to systems designed for modelling, analysis, and visualisation of continuous fields. Because it is impossible to cover all or even most of the existing interpolation techniques, only methods which are often used in connection with GIS or have the potential to be widely used for GIS applications are included, and references are given to literature for more detailed descriptions.

Local Neighbourhood Approach

Local methods are based on the assumption that each point influences the resulting surface only up to a certain finite distance. Values at different un-sampled points are computed by functions with different parameters, and the condition of continuity between these functions is defined only for some approaches. The method of point selection used for the computation of the interpolating function differs among the various methods and their concrete implementations.

Inverse distance weighted interpolation (IDW): This is one of the simplest and most readily available methods. It is based on an assumption that the value at un-sampled point can be approximated as a weighted average of values at points within a certain cut-off distance, or from a given number m of the closest points (typically 10 to 30). Weights are usually inversely proportional to a power of distance which, at an un sampled location r, leads to an estimator,

$$F(r) = \sum_{i=1}^{m} w_i z(r_i) = \frac{\left(\sum_{i=1}^{m} z(r_i)/|r-r_i|^p\right)}{\sum_{j=1}^{m} 1/|r-r_j|^p}$$

where, p is a parameter (typically p=2; for more details on the influence of this parameter). While this basic method is easy to implement and is available in almost any GIS, it has some well-known shortcomings that limit its practical applications. The method often does not reproduce the local shape implied by data and produces local extreme at the data points. A number of enhancements has been suggested, leading to a class of multivariate blended IDW surfaces and volumes. However, most of these modifications are not implemented within GIS.

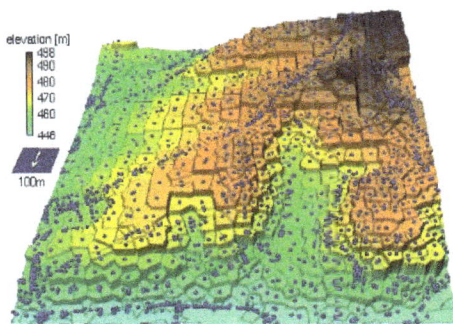

(a)

(b)

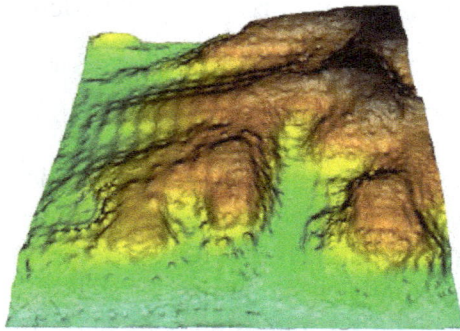

(c)

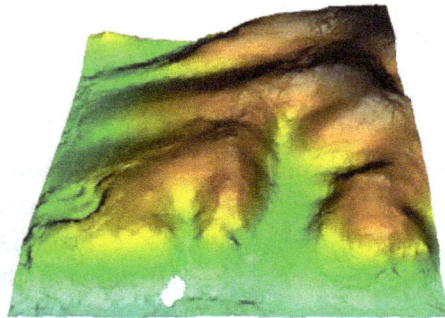

(d)

(e)

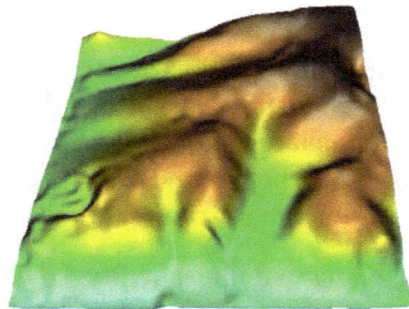

(f)

Natural Neighbour Interpolation

This uses a weighted average of local data based on the concept of natural neighbour coordinates derived from Thiessen polygons for the bivariate, and Thiessen poly hedra for the trivariate case. The value in an un-sampled location is computed as a weighted average of the nearest neighbour values with weights dependent on areas or volumes rather than distances. The number of given points used for the computation at each un-sampled point is variable, dependent on the spatial configuration of data points. Natural neighbour linear interpolation leads to a rubber-sheet character of the resulting surface. The addition of blended gradient information (derived from data points by local 'pre-interpolation') allows the surface to be made smooth everywhere with tautness, analogous to tension, tuned according to the character of the modelled phenomenon. The value of tautness is controlled by two empirically selected parameters which modify the shape of the blending function. The result is a surface with smoothly changing gradients and passing through data points, blended from natural neighbour local trends, with local tune able tautness, and with the capability to calculate derivatives and integrals. The method has been used typically for topographic, bathymetric, geophysical, and soil data.

Interpolation Based on a Triangulated Irregular Network (TIN)

This uses a triangular tessellation of the given point data to derive a bivariate function for each triangle which is then used to estimate the values at un-sampled locations. Linear interpolation uses planar facets fitted to each triangle (Non-linear blended functions (e.g. polynomials) use additional continuity conditions in first-order, or both first- and second-order derivatives (C1, C2), ensuring smooth connection of triangles and differentiability of the resulting surface . Because of their local nature, the methods are usually fast, with an easy incorporation of discontinuities and structural features.

Appropriate triangulation respecting the surface geometry is crucial. Extension to d-dimensional problems is more complex than for the distance based methods. While a TIN provides an effective representation of surfaces useful for various applications, such as dynamic visualisation and visibility analyses, interpolation based on a TIN, especially the simplest, most common linear version, belongs among the least accurate methods.

Rectangle-based Methods

These are analogons to a TIN and involve fitting blended polynomial functions to regular or irregular rectangles, such as Hermite, Bezier, or B-spline patches, often with locally tunable tension. These methods were developed for computer-aided design and computer graphics and are not very common in GIS applications.

Spatial Regression

Spatial data exhibits two properties that make it difficult (but not impossible) to meet the assumptions and requirements of traditional (non-spatial) statistical methods, like OLS regression:

- Geographic features are more often than not spatially auto correlated; this means that features near each other tend to be more similar than features that are farther away. This creates an over count type of bias for traditional (non-spatial) regression methods.

- Geography is important, and often the processes most important to what you are modelling are non-stationary; these processes behave differently in different parts of the study area. This characteristic of spatial data can be referred to as regional variation or nonstationarity.

True spatial regression methods were developed to robustly manage these two characteristics of spatial data and even to incorporate these special qualities of spatial data to improve their ability to model data relationships. Some spatial regression methods deal effectively with the first characteristic (spatial autocorrelation), others deal effectively with the second (nonstationarity).

Spatial Interaction

One methodology of particular importance to transport geography relates to how to estimate flows between locations, since these flows, known as spatial interactions, enable to evaluate the demand (existing or potential) for transport services. They cover a wide variety of movements such as journeys to work, migrations, tourism, the usage of public facilities, the transmission of information or capital, the market areas of retailing activities, international trade and freight distribution. These movements can be physical (people or freight) or intangible (information).

Economic activities are generating (supply) and attracting (demand) flows. The simple fact that a movement occurs between an origin and a destination underlines that the costs incurred by a spatial interaction are lower than the benefits derived from such an interaction. As such, a commuter is willing to drive one hour because this interaction is linked to an income, while international trade concepts, such as comparative advantages, underline the benefits of specialization and the ensuing generation of trade flows between distant locations.

Three interdependent conditions are necessary for a spatial interaction to occur:

- Complementarity: There must be a supply and a demand between the interacting locations. A residential zone is complementary to an industrial zone because the first is supplying workers while the second is supplying jobs. The same can be said concerning the complementarity between a store and its customers and between an industry and its suppliers (movements of freight).

- Intervening opportunity (lack of): Refers to a location that may offer a better alternative as a point of origin or as a point of destination. For instance, in order to have an interaction of a customer to a store, there must not be a closer store that offers a similar array of goods.

- Transferability: Freight, persons or information being transferred must be supported by transport infrastructures, implying that the origin and the destination must be linked. Costs to overcome distance must not be higher than the benefits of related interaction, even if there is complementarity and no alternative opportunity.

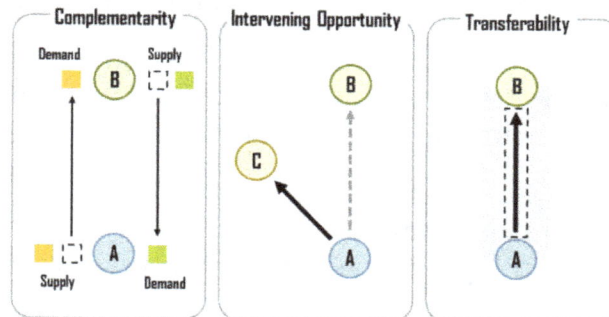

Conditions for the Realization of a Spatial Interaction

Spatial interaction models seek explain existing spatial flows. As such it is possible to measure flows and predict the consequences of changes in the conditions generating them. When such attributes are known, it is possible, for example, to better allocate transport resources such as highways, buses, airplanes or ships since they would reflect the expected transport demand more closely.

Origin / Destination Matrices

Each spatial interaction, as an analogy for a set of movements, is composed of a discrete origin / destination pair. Each pair can itself be represented as a cell in a matrix where rows are related to the locations (centroids) of origin, while columns are related to locations (centroids) of destination. Such a matrix is commonly known as an origin / destination matrix, or a spatial interaction matrix.

O/D Pair		O/D Matrix			
		Destinations			
		A	B	C	Total
Origins	A				Ti
	B				
	C				
	Total	Tj			T

In the O/D matrix, the sum of a row (Ti) represents the total outputs of a location (flows originating from), while the sum of a column (Tj) represents the total inputs (flows bound to) of a location. The summation of inputs is always equals to the summation of outputs. Otherwise, there are movements that are coming from or going to outside the considered system. The sum of inputs or outputs gives the total flows taking place within the system (T). It is also possible to have O/D matrices according to the age group, income, gender, etc. Under such circumstances they are labelled sub-matrices since they account for only a share of the total flows. If the sample is small and disaggregated it is possible to use a simple list of interactions instead of a matrix. Still, an origin / destination matrix can be constructed out of this list.

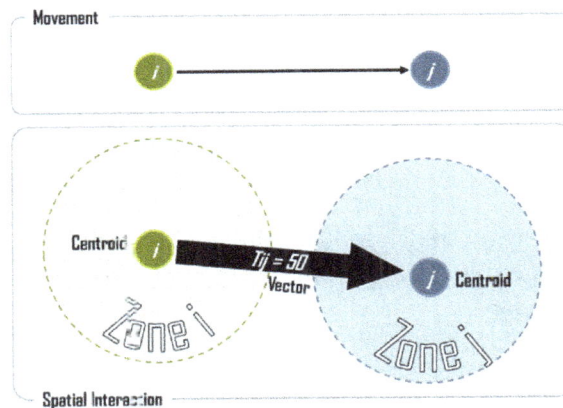

Representation of a Movement as a Spatial Interaction

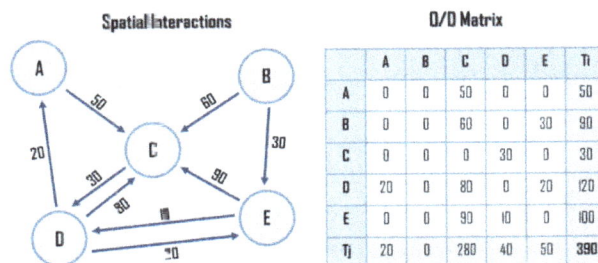

	A	B	C	D	E	Ti
A	0	0	50	0	0	50
B	0	0	60	0	30	90
C	0	0	0	30	0	30
D	20	0	80	0	20	120
E	0	0	90	10	0	100
Tj	20	0	280	40	50	390

Constructing an O/D Matrix

In many cases where spatial interactions information is relied on for planning and allocation purposes, origin / destination matrices are not available or are incomplete. Palliating this lack of data commonly requires surveys. With economic development, the addition of new activities and transport infrastructures, spatial interactions have a tendency to change very rapidly as flows adapt to a new spatial structure. The problem is that an origin / destination survey is very expensive in terms of efforts, time and costs. In a complex spatial system such as a region, O/D matrices tend to be quite large. For instance, the consideration of 100 origins and 100 destinations would imply 10,000 separate O/D pairs for which information has to be provided. In addition, the data gathered by spatial interaction surveys is likely to rapidly become obsolete as economic and spatial conditions change. It is therefore important to find a way to estimate as precisely as possible spatial interactions, particularly when empirical data is lacking or is incomplete. Further, the emergence of 'big data' has enabled to collect large amounts of personal mobility information that is possible to convert into flows between spatial units. A possible solution relies on using a spatial interaction model to complement and even replace empirical observations.

Spatial Interaction Models

Spatial interaction models are usually the first two steps in the standard four step transportation/land use model as they estimate the spatial generation and distribution of trips. The basic assumption concerning many spatial interaction models is that flows are a function of the attributes of the locations of origin, the attributes of the locations of destination and the friction of distance between the concerned origins and the destinations. The general formulation of the spatial interaction model is as follows:

$$T_{ij} = f(V_i, W_j, S_{ij})$$

- T_{ij}: Interaction between location i (origin) and location j (destination). Its units of measurement are varied and can involve people, tons of freight, traffic volume, etc. It also relates to a time period such as interactions by the hour, day, month or year.

- V_i: Attributes of the location of origin i. Variables often used to express these attributes are socio-economic in nature, such as population, number of jobs available, industrial output or gross domestic product.

- W_j: Attributes of the location of destination j. It uses similar socio-economic variables than the previous attribute to underline the reciprocity of the locations.

- S_{ij}: Attributes of separation between the location of origin i and the location of destination j. Also known as transport friction, friction of distance or impedance. Variables often used to express these attributes are distance, transport costs or travel time.

The attributes of V and W tend to be paired to express complementarity in the best possible way. For instance, measuring commuting flows (work-related movements) between different locations would likely consider a variable such as working age population as V and total employment as W. From this general formulation, three basic types of interaction models can be constructed:

- Gravity model: Measures interactions between all the possible location pairs.

- Potential model: Measures interactions between one location and every other location.

- Retail model: Measure the boundary of the market areas between two locations competing over the same market based upon the intensity of their respective interactions.

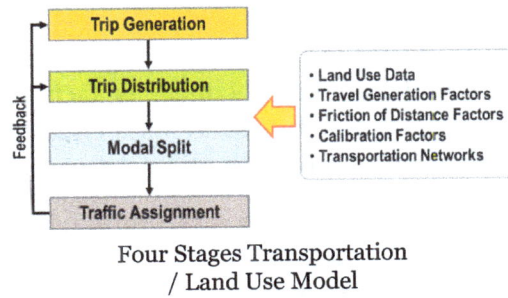

Four Stages Transportation
/ Land Use Model

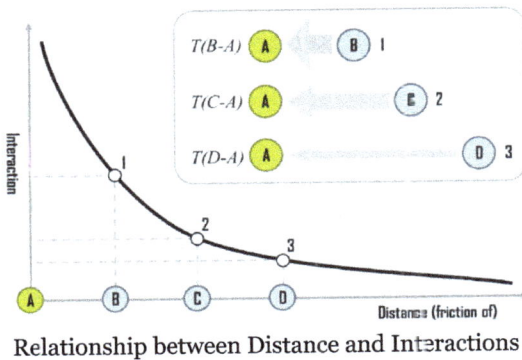

Relationship between Distance and Interactions

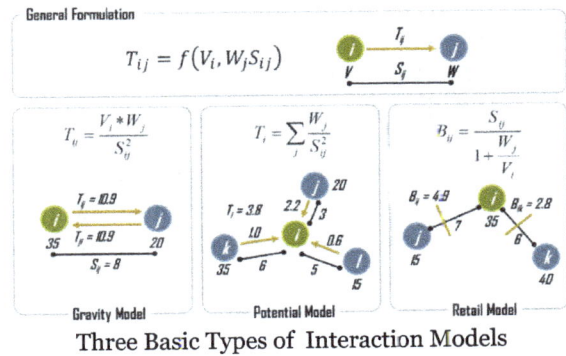

Three Basic Types of Interaction Models

The Gravity Model

The gravity model is the most common formulation of the spatial interaction method. It is named as such because it uses a similar formulation than Newton's law of gravity. Gravity like representations have been applied in a wide variety of contexts, such as migration, commodity flows, traffic flows, commuting, and evaluating boundaries between market areas. Accordingly, the attraction between two objects is proportional to their mass and inversely proportional to their respective distance. Consequently, the general formulation of spatial interactions can be adapted to reflect this basic assumption to form the elementary formulation of the gravity model:

$$T_{ij} = k\frac{P_i P_j}{d_{ij}}$$

- P_i and P_j: Importance of the location of origin and the location of destination.

- d_{ij}: Distance between the location of origin and then location of destination.

- K: proportionality constant related to the rate of the event. For instance, if the same system of spatial interactions is considered, the value of k will be higher if interactions were considered for a year comparatively to the value of k for one week.

Thus, spatial interactions between locations i and j are proportional to their respective importance divided by their distance. The gravity model can be extended to include several calibration parameters:

$$T_{ij} = k\frac{P_i^\lambda P_j^\alpha}{d_{ij}^\beta}$$

- P, d and k refers to the variables previously discussed.

- β (beta): A parameter of transport friction related to the efficiency of the transport system between two locations. This friction is rarely linear as the further the movement the greater the friction of distance. For instance, two locations services by a highway will have a lower beta index than if they were serviced by a regular road.

- λ (lambda): Potential to generate movements (emissivity). For movements of people, lambda is often related to an overall level of welfare. For instance, it is logical to infer that for retailing flows, a location having higher income levels will generate more movements (customers).

- α (alpha): Potential to attract movements (attractiveness). Related to the nature of economic activities at the destination. For instance, a centre having important commercial activities will attract more movements.

A significant challenge related to the usage of spatial interaction models, notably the gravity model, is related to their calibration. Calibration consists in finding the value of each parameters of the model (constants and exponents) to insure that the estimated results are similar to the observed flows and that those results can be replicated. If it is not the case, the model is of limited use as it predicts or explains little. It is impossible to know if the process of calibration is accurate without comparing estimated results with empirical evidence. Consistent calibration enables the model to be more rigorous and adaptable to other contexts.

In the two formulations of the gravity model that have been introduced, the simple formulation offers a good flexibility for calibration since four parameters can be modified. Altering the value of beta, alpha and lambda will influence the estimated spatial interactions. Furthermore, the value of the parameters can change in time due to factors such as technological innovations, new transport infrastructure and economic development. For instance, improvements in transport efficiency generally have the consequence of reducing the value of the beta exponent (friction of distance). Economic development is likely to influence the values of alpha and lambda, reflecting a growth in mobility.

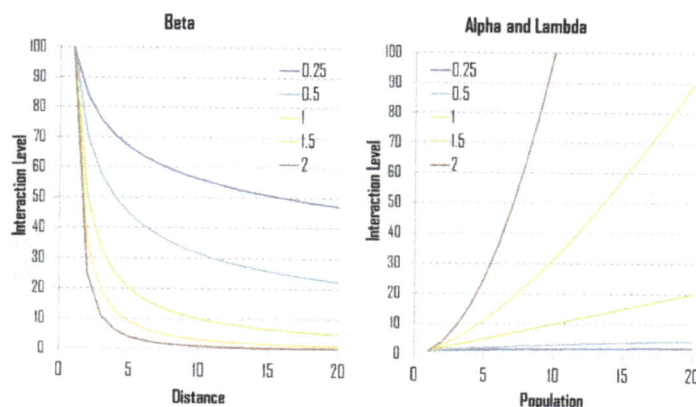

Effects of beta, alpha and lambda on Spatial Interactions

Calibration can also be considered for different O/D matrices according to age, income, gender, type of merchandise and modal choice. A part of the scientific research in transport and regional planning aims at finding accurate parameters for spatial interaction models. This is generally a costly and time consuming process, but a very useful one. Once a spatial interaction model has been validated for a city or a region, it can then be used for simulation and prediction purposes, such as how many additional flows would be generated if the population increased or if better transport infrastructures (lower friction of distance) were provided.

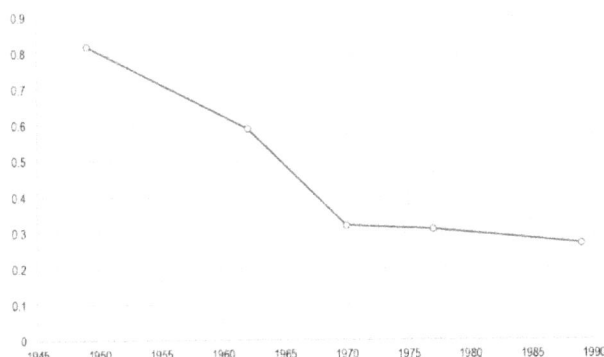

Chicago's beta Values for Air Transportation, 1949-1989

Outside the gravity model, there are other models that can be used to measure spatial interactions. Destination choice models are considered an extension of the gravity model that is gaining popularity since it provides a wider range of factors explaining the assignment of spatial interactions. While the gravity model assumes that flows are generated as a function of attributes of the origin and destination pondered by impedance functions, the destination choice model allows for additional behavioural attributes to mobility, including income, walkability, the availability of parking and psychological barriers. The main goal is to explain flows that the standard gravity model does not capture well.

References

- Spatial-autocorrelation-moran-i-gis: gisgeography.com, Retrieved 02 July 2018
- Spatial-heterogeneity: gispopsci.org, Retrieved 21 June 2018
- Gis-stratified-heterogeneity: gislounge.com, Retrieved 11 June 2018
- Spatial-analysis-interpolation, gentle-is-introduction: docs.qgis.org, Retrieved 25 May 2018
- Regression-analysis-basics, spatial-statistics-toolbox: desktop.arcgis.com, Retrieved 13 April 2018

Geospatial Analysis

Geospatial analysis refers to the approach that involves the application of statistical analysis and different analytical techniques to geographical or spatial data. The aim of this chapter is to explore the fundamental concepts of geospatial analysis, such as surface analysis, network analysis and geovisualization which will closely aid in its understanding.

Geospatial data that is data with location information is generated in huge volumes by billions of mobile phones, sensors, and other sources every day. Data begets data, constantly ratcheting up the unbounded streams of geospatial data ("geo data" for short) awaiting our analysis. Where people and their machines go, what our remote sensing detects, and how our devices are arrayed in space and perform across time all matter a great deal to the vibrancy of our economy, the health of our planet, and our general happiness and well-being. Geospatial analytics can provide us with the tools and methods we need to make sense of all that data and put it to use in solving problems we face at all scales.

Geospatial analytics has its contextual roots in print cartography and its contemporary development in defense. In the United States, the Department of Defense (DoD) has traditionally been the biggest consumer of geospatial analytics. Intelligence community satellites have been producing constant streams of telemetry for decades now, and defense vehicles, sensor nets, and many other sources of data have sprung up over time. With the rise of this type of data, the DoD has helped promote open source, open data, and data analysis companies such as Socrata, Data bricks, and Uncharted Software. Much could be written (and indeed has) on the history of geographic information systems (GISs) and geospatial work for defense; rather than rehashing that here, we recommend referring to existing resources for more background, including the history of the Open Geospatial Consortium (OGC), which originated as a voluntary consensus standards organization thanks to efforts from the US Army Corps of Engineers and a great many others in public and private enterprises.

It is important to note that geospatial data is used far beyond the defense realm, however. Consider that just about any local, state, or federal government agency is responsible for a variety of geospatial analytics problems. For example, something as seemingly straightforward as a pothole in a road may require one road crew truck to drive by simply to confirm the need for repair, followed by another truck to actually perform the work. The locations of all those potholes and trucks need to be tracked. This example is multiplied many times over when you consider that there are similar geospatial requirements for crime reporting, traffic collision monitoring and assistance, building permitting, and a great many other governmental functions. Likewise, in geo focused industries like real estate, cadastral mapping, land records, and property exchange archives inform a practice of land use commercialization grounded in geospatial information.

Beyond governmental use cases (satellite telemetry, road potholes, Ships at sea, etc.), another driver for geo analysis in Big Data has been telecom. Visit the Ericsson headquarters north of Stockholm, and you'll find the Swedish headquarters for Esri, a supplier of GIS software and applications, directly next door. Mobile operators in particular generate massive amounts of streaming data—telemetry about whether their equipment in the field is working correctly, whether their subscribers will be filing service complaints, and so on—much of which requires geospatial analysis.

But it's not just government and mobile operators that are generating and capturing valuable geospatial data. For starters, anyone who has a fleet of trucks faces a range of geospatial analytics problems; supply chain management, routing efficiency and resourcing of people, products, and vehicles all operate under an umbrella of geospatial influence factors. This opens new opportunities for leveraging geospatial Big Data in mainstream business.

From satellite and cell phone data, we can glean information about the areas hardest hit by a natural disaster as it occurs, and determine the places and people that most urgently need assistance. Similarly, satellite and remote sensing data can be used to analyze and ultimately fight the effects of climate change, shedding light on both short-term mitigation efforts and long-term solutions to emission problems.

From ambient noise sensors distributed around municipalities and perhaps even from apps running on mobile phones and other personal devices, we might be able to monitor and discover useful information on a cityscape's noise levels. This information could be used to report potentially dangerous noise from air traffic, construction projects, or highways and busy streets, and these reports could in turn be used to enforce noise ordinances and safer, more pleasant city zoning and design.

Non-profits can use data about their donors and communities to better plan and communicate the services they provide. We can also proactively improve public health by analyzing an individual's fitness tracking data, comparing it to norms for cohorts, and giving suggestions or "nudges" about behavioral tweaks and habits that could lead to large improvements in health and fitness down the road.

With the rise of the Internet of Things, IoT sensor networks are pushing the geospatial data rates even higher. There has been an explosion of sensor networks on the ground, mobile devices carried by people or mounted on vehicles, drones flying overhead, tethered aerostats (such as Google's Project Loon), atmosats at high altitude, and microsats in orbit.

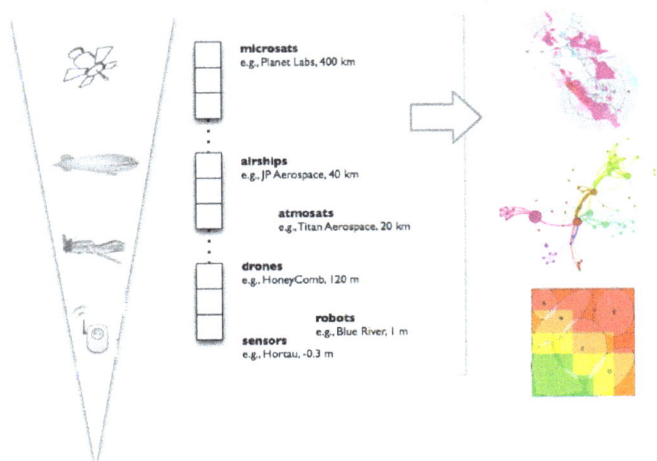

microsats
e.g., Planet Labs, 400 km

airships
e.g., JP Aerospace, 40 km

atmosats
e.g., Titan Aerospace, 20 km

drones
e.g., HoneyComb, 120 m

robots
e.g., Blue River, 1 m

sensors
e.g., Hortau, -0.3 m

Figure: Layers of data sources

These layered sensing networks—layers of data sources, ranging from sensors on the ground to vehicles and mobile devices to drones to satellites, all at complementary levels of detail and cost/performance—serve as a flywheel for a tremendous spike in geospatial data Ground sensors can be relatively expensive, but they provide detail, whereas satellites provide the broader perspective less expensively at scale over time.

The bottom line here is most often about control systems. Some "thing" needs to be automated, and that automation process needs to be optimized. If that thing moves within the world, it generates geo data and must leverage spatial analytics. That holds for drones, self-driving cars, and robots in general—and in the larger picture, even companies like Uber can be considered as using complex control systems.

New Solutions for More and Complex Geospatial Data

Whatever problems we face, whether at the local, regional, national, or even global scale, if they involve a "where" component, geospatial solutions can probably be brought to bear to improve the result. It may be unclear at the onset of tackling a geospatial analytics challenge, however, which solution would be best. For many tasks, the data is not truly "big," and in fact mainstream solutions and commodity hardware may prove sufficient to address them. For others, Big Data techniques are required.

For a long while the mainstay of GIS data systems has been the ArcGIS product line at Esri Visit just about any municipal, state, or federal agency tracking geo tagged items, and you'll find a number of ArcGIS subscriptions—which tends to make Esri seem like the proverbial 800-pound gorilla in the room. Historically much of that industry dominance involves compatibility with proprietary desktop software (Microsoft Windows) and relatively modest dataset sizes, perhaps in the megabyte or gigabyte rang.

Whether using ArcGIS or other tools, geospatial work requires atypical data types (e.g., points, shape files, map projections), potentially many layers of detail to process and visualize, and specialized algorithms—not your typical ETL (extract, transform, load) or reporting work. A sample of the complexities in many geospatial analyses might include the following:

- At their foundation, most geospatial applications require some kind of map. Tiling provides rectangles for a selected level of detail, generally raster graphics.

- Analytics overlays, such as vector graphics, can be layered atop the tiling.

- Data sources may be relatively sparse and require statistical smoothing or interpolation (e.g. kriging to convert discrete data points into heat maps, choropleths, and so on that are more useful to visualize data as geospatial overlays).

- Some data sources (e.g., satellite images) have inherent needle-in-a-haystack problems that require sophisticated algorithms to identify points of interest, or locations that change dramatically over time (e.g., a building under construction).

- Other data sources—for example, business addresses—provide metadata for maps but may have conflicting information to resolve (e.g., multiple addresses for a business).

- Data sources come in a bewildering number of formats. This is a hard problem. If you thought that JSON versus Thrift versus Avro versus Parquet versus ORC File was complex, brace yourself for the complexities of geo data! You can see many examples by perusing the list of OGR vector formats or reading through documentation for libraries such as GDAL (the Geospatial Data Abstraction Library), which supports many raster and vector formats to abstract away format-related complexities for you.

- Meanwhile, map tiles, data sources, analytics, and the like may bring in a variety of licensing issues and conflicts.

- Once you have the tiles, the data sources, the metadata, and the analytics, you need an interactive platform for zooming, selecting points, selecting optional layers and more.

- Then comes the part that requires real expertise: design, data visualization, interpretation and story-telling.

Surface Analysis

Surface analysis is a type of investigation performed on the outermost layer of a solid material. It specifically refers to an analysis of the thin layer of the solid that is in contact with its environment. Surface analysis is performed by exciting the surface of a material using an energy source, and then observing and analyzing what occurs as a result of this energization to determine the properties of a material's surface.

Surface analysis involves several kinds of processing, including extracting new surfaces from existing surfaces, reclassifying surfaces, and combining surfaces. Certain tools extract or derive information from a surface, a combination of surfaces, or surfaces and vector data.

Terrain Analysis Tools

Some of these tools are primarily designed for the analysis of raster terrain surfaces. These include Slope, Aspect, Hill shade, and Curvature tools.

Below is an example of an elevation raster in planimetric and perspective views.

Elevation

High : 4361

Low: 438

The Slope tool calculates the maximum rate of change from a cell to its neighbors, which is typically used to indicate the steepness of terrain.

Below is an example of Slope raster in planimetric and perspective views.

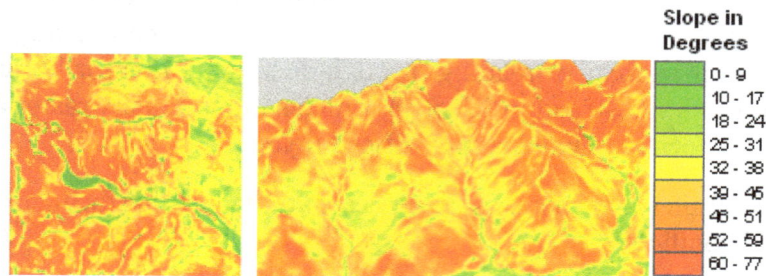

Slope raster derived from elevation Slope map legend.

The Aspect tool calculates the direction in which the plane fitted to the slope faces for each cell. The aspect of a surface typically affects the amount of sunlight it receives (as does the slope); in northern latitudes places with a southerly aspect tends to be warmer and drier than places that have a northerly aspect.

Below is an example of an aspect raster in planimetric and perspective views.

Hill shade shows the intensity of lighting on a surface given a light source at a particular location; it can model which parts of a surface would be shadowed by other parts.

Below is an example of a hill shade raster in planimetric and perspective views.

Curvature calculates the slope of the slope (the second derivative of the surface), that is, whether a given part of a surface is convex or concave. Convex parts of surfaces, like ridges, are generally exposed and drain to other areas. Concave parts of surfaces, like channels, are generally more sheltered and accept drainage from other areas. The Curvature tool has a couple of optional variants, Plan and Profile Curvature. These are used primarily to interpret the effect of terrain on water flow and erosion. The profile curvature affects the acceleration and deceleration of flow, which influence erosion and deposition. The planiform curvature influences convergence and divergence of flow.

Below is an example of a curvature raster in planimetric and perspective views.

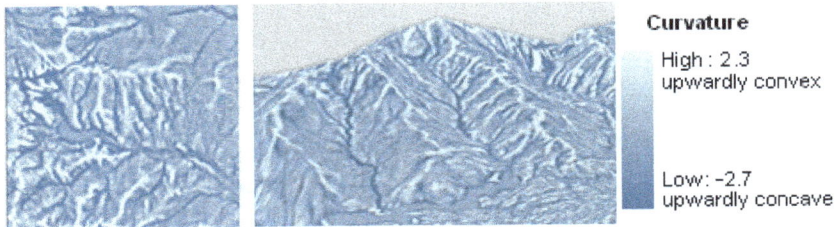

Curvature

High : 2.3
upwardly convex

Low: –2.7
upwardly concave

Visibility Tools

Some tools are used to analyze the visibility of parts of surfaces. The Line of Sight tool identifies whether or not one location is visible from another and whether or not the intervening locations along a line between the two locations are visible.

Below is an example of a Line of Sight analysis. An observer at the southern end of the line can see the parts of the terrain along the line that are colored green, and cannot see the parts of the terrain along the line that are colored red. In this case, the observer cannot see the fire in the valley on the other side of the mountain.

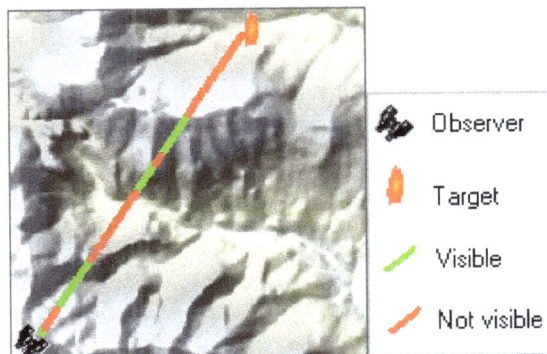

Observer

Target

Visible

Not visible

The visibility tools support offsets, which allow you to specify the height of the observer points and the observed points or cells.

Below is an example of a Line of Sight analysis comparing the results with no offset and with a target offset. Locations along the line that are visible to the observer are green, those that are hidden by intervening terrain are red.

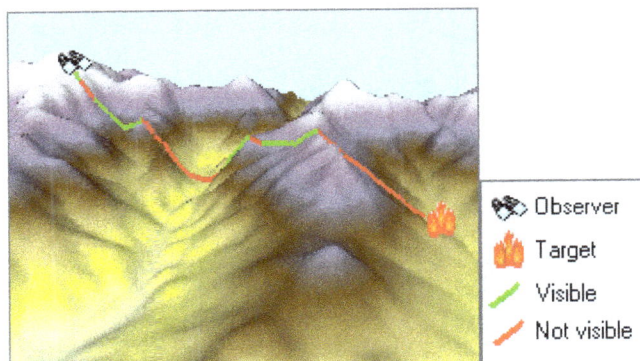

Observer

Target

Visible

Not visible

You might use a target offset to model a building or a smoke plume.

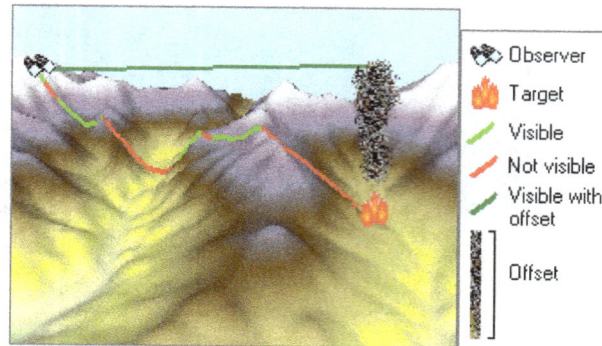

With a large target offset the target is visible, even though the visibility of the points along the intervening terrain do not change.

You could add an offset to the observer as well, to model a tower at the observer location. Adding an observer offset generally increases the amount of terrain that is visible from a location.

The Observer Points tool identifies which observers, specified as a set of points, can see any given cell of a raster surface. The View shed tool calculates for each cell of a raster surface and a set of input points (or the vertices of input lines) how many observers can see any given cell.

Below is an example of a View shed analysis with a single input observer point. The observer has an offset to model the view from a fire tower 50 meters taller than the ground surface. Cells outside of the observer's view shed are blacked out in the image on the right.

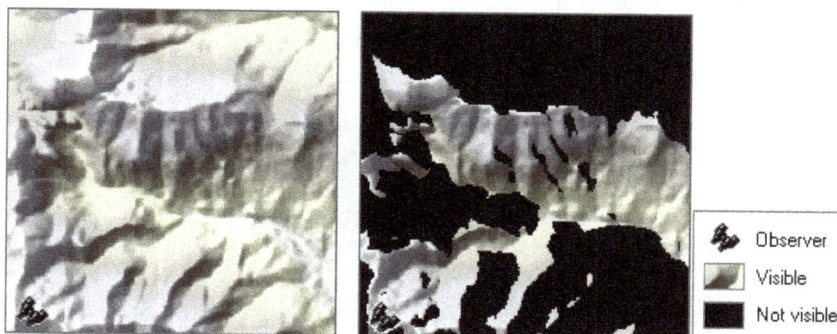

In the perspective views below you can see the observer point and the terrain.

Ridges hide the valleys behind them from the observer point.

Both the Observer Points and View shed tools also allow you to specify observer and target offsets, as well as a set of parameters that let you limit the directions and distance that each observer can view.

Volume Tools

Some tools are used to calculate volumes from surface information. These tools calculate the difference in volume between a raster or TIN surface and another surface. Depending on the tool, the other surface might be specified by a horizontal plane at a given elevation or by a second raster or TIN surface.

Below is an example of a terrain surface representing the typical fill level of a reservoir You could use the volume tools to calculate the volume of additional water when the reservoir is near capacity.

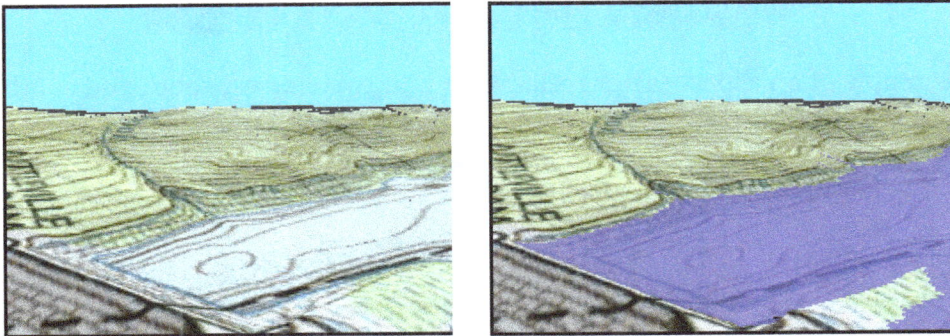

The Surface Volume tool is used to calculate volume of a surface above or below a horizontal plane at a specific elevation. You might use this tool to calculate the volume of water in a section of river channel at a particular flood stage. This tool can be used on raster or TIN surfaces. The output of the tool is a text file reporting the parameters used and the resulting surface area and volumes.

The Cut/Fill tool is used to calculate the amount of difference in each cell for a before and after raster of the same area. This tool could be used to calculate the volume of earth that must be brought to or removed from a construction site to reshape a surface. This tool works on two raster and the results are presented as a raster of the difference between the two layers.

The TIN Difference tool is similar to the Cut/Fill tool, but it works on a pair of input TIN surfaces. This tool creates a polygon feature class where each polygon is given attributes identifying whether the second TIN is above, below, or the same as the first TIN, and the volume of the difference between the TINs in that polygon.

The TIN Polygon Volume tool calculates the volume difference and surface area for each polygon in a feature class relative to a TIN surface. Each polygon in the feature class represents a horizontal area at an elevation specified in a height field. The volume above or below this planar area to the TIN surface is added to a volume field in the feature class, and the surface area of the polygon is added to a surface area field.

Reclassification Tools

One way to convert surface data into more usable information for an analysis is to reclassify the surface. Reclassifying a surface sets a range of values equal to a single value. You might reclassify a surface so that areas with cells above a given value, or between two critical values, are given one code, and other areas are given another; or, you might use the reclassify (or slice) tool to divide a surface up into a given number of classes as a means of aggregating and generalizing detailed data. Reclassifying surfaces is often done to reduce the number of output categories for an overlay analysis.

Below is an example of an elevation raster sliced into several classes (each class represents a range of elevation values) and reclassified into two classes (above and below a given elevation).

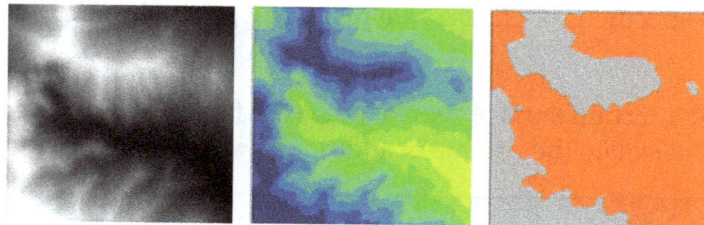

Below is an example of an aspect raster reclassified into two classes; south and southwest aspect slopes have a value of 1 (light) and other aspects have a value of 0 (dark).

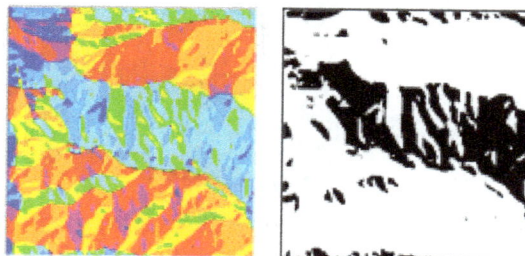

Distance Tools

Some distance tools create raster that show the distance of each cell from a set of locations.

The tools include the shortest straight-line distance to a set of source features, and the direction of the closest feature. The Euclidean Allocation tool creates zones of a surface that are allocated to the closest feature.

The Cost Distance, Cost Path, Cost Back Link, and Cost Allocation tools are used to find the shortest (least cost) path from sources to destinations, taking into account a raster that quantifies the cost of traversing the surface. The cost raster may reflect difficulty, energy, time, or dollar costs, or a unit less composite of several factors that influence the cost of travel or flow across a surface.

The Path set of tools perform much the same function as the Cost set, but take the additional factors of surface distance and vertical travel difficulty (cost) into account; that is, the fact that the length of a given line over hilly terrain is longer than the same line on a perfectly flat surface, and the fact that it may be easier to move along a slope than it is to move up or down the slope.

Overlay Tools

Raster overlay tools combine two or more raster using logical, arithmetic, or weighted combination methods.

Map Algebra allows you to combine surfaces using logical or arithmetic operators. The Weighted Overlay and Weighted Sum tools allows you to combine multiple raster of varying importance. This is useful in site suitability analyses when several factors contribute to suitability, but certain factors contribute more heavily than others.

Some tools perform algebraic or logical operations upon surfaces. The Spatial Analyst Neighbourhood functions, such as the block and focal functions, compute values for the cells of an output raster based on the values of surrounding cells; these can be used to remove noise or enhance edge contrasts, or resample raster to a lower resolution. Local functions combine, compare, or summarize several raster on a cell by cell basis. Zonal functions calculate for each cell some function or statistic using the value for all cells belonging to the same zone.

Extracting Information from Surfaces

Some tools extract vector features from surfaces, or produce tabular summaries or smaller raster samples of surfaces.

Sampling Rasters

The Sample tool creates a table that shows the values of a raster, or several rasters, at a set of sample point locations. The points can be in a point feature class or the cells in a raster that have values other than No Data. You might use this tool to get information about what occurs at a set of points, such as bird nesting sites, from terrain, distance to water, and forest type rasters.

Below is an example of a geology raster being sampled at a set of points; the result is a table.

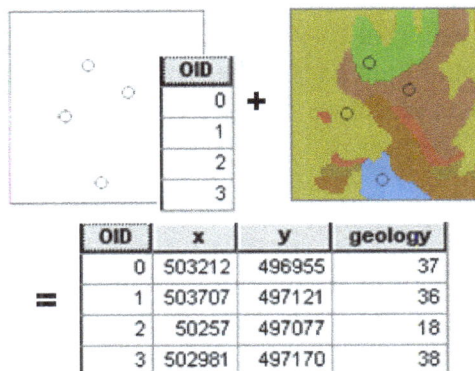

OID	x	y	geology
0	503212	496955	37
1	503707	497121	36
2	50257	497077	18
3	502981	497170	38

The output table can be analysed on its own, or joined to the sample point features.

Below is an example of the sample results table joined back to the original sample points.

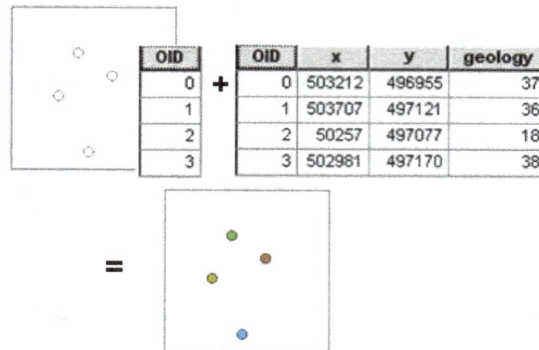

The Extract tools create a new raster with a copy of the cells within some mask area. The Extract by Mask tool lets you use a polygon feature class to extract the raster data.

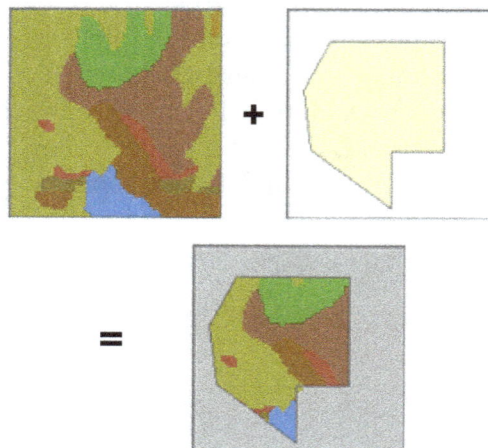

The Extract Values to Points tool creates a new feature class of points with the values of a single raster at a set of input point features. The Extract by Attributes tool selects cells of a raster based on a logical query. Extract by Polygon and Extract by Rectangle take lists of coordinate values that define an area and output a raster that is either inside or outside of the polygon. Extract by Circle takes the centre coordinates and radius of a circle and outputs a raster that is either inside or outside of the circle. Extract by Points takes a list of coordinate values that define a set of points and outputs a raster of the cell values at these points (or excluding these points). In all cases, the cells from the original raster that are not part of the Extract area are given No Data values. The 3D Analyst Surface Spot tool extracts elevation values from a surface for a set of point features and adds them to a Spot attribute of the points.

Extracting Information from a TIN

TINs store slope and aspect information as attributes of the TIN facets. Rather than deriving slope and aspect for TIN surfaces (as you do with raster terrain models, which only store the elevation values) you simply need to extract that information from the facets to a set of polygons. TIN Aspect and TIN Slope extract aspect and slope data from a TIN and add that information as attributes of a polygon feature class.

Below is an example of TIN elevation model and the aspect information it contains.

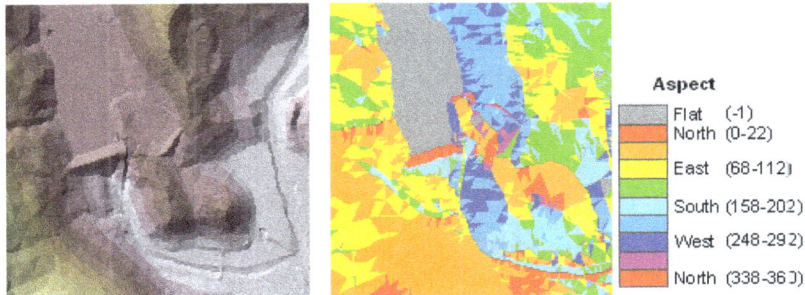

Below is an example of TIN elevation model and the slope information it contains.

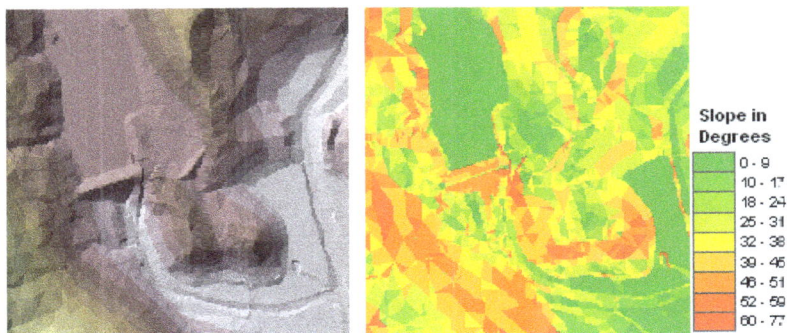

Extracting Contours

The Contour tool extracts lines of constant value (isolines) from a raster surface. The TIN Contour tool extracts a line feature class of contours from a TIN surface.

Below is an example of an elevation model and contour lines extracted from it.

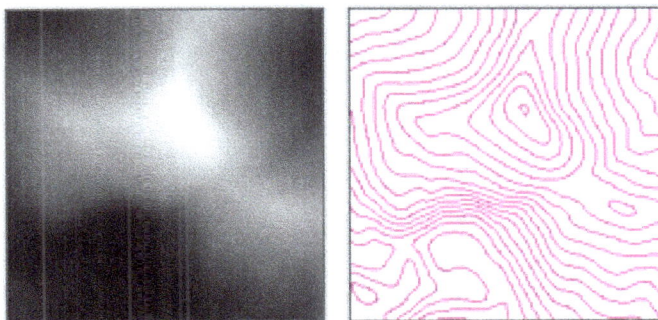

Zonal statistics tools can produce tables of summary statistics for a given raster, based on zones defined by another raster or a polygon feature class, or it can produce a new raster that corresponds to the zones with a specific summary statistic as an attribute.

Hydrology Tools

Hydrology tools derive drainage basin and stream information from terrain rasters; this information can be converted to vector features. The process requires several tools that derive information from the terrain surface, resulting in basin and stream rasters that can be converted to vector features. The Flow Direction tool takes a terrain surface and identifies the down-slope direction for each cell. The Basin tool uses the results of the Flow Direction tool to identify the drainage basins, made up of the connected cells that drain to a common location. The Flow Accumulation tool identifies how much surface flow accumulates in each cell; cells with high accumulation values are usually stream or river channels. It also identifies local topographic highs (areas of zero flow accumulation) such as mountain peaks and ridgelines.

Below is an example of an elevation model:

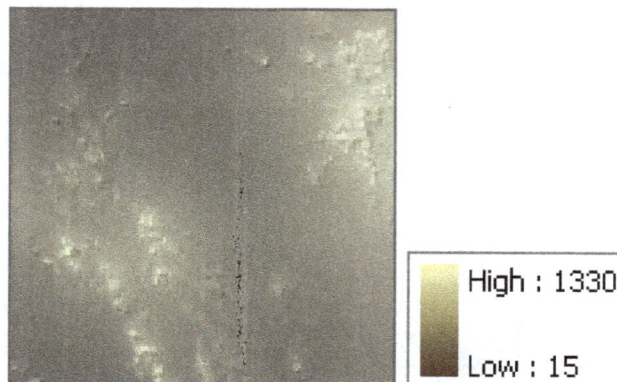

Below is an example of a flow direction surface derived from the elevation model:

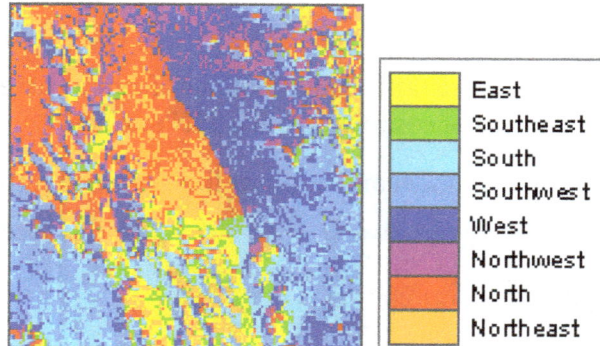

Below is an example of basins derived from the flow direction surface:

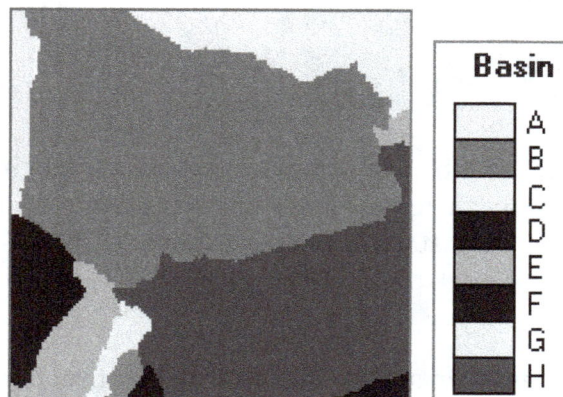

Below is an example of a flow accumulation surface derived from the flow direction surface:

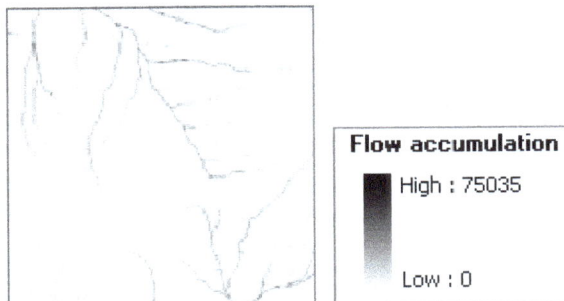

The flow accumulation surface can be processed with a Map Algebra Conditional (Con) statement such as:

    con (flowacc > 100, 1)

to capture only those cells with high flow accumulation values (in this case, greater than 100) into a stream raster.

Below is an example of a stream raster extracted from the flow accumulation surface.

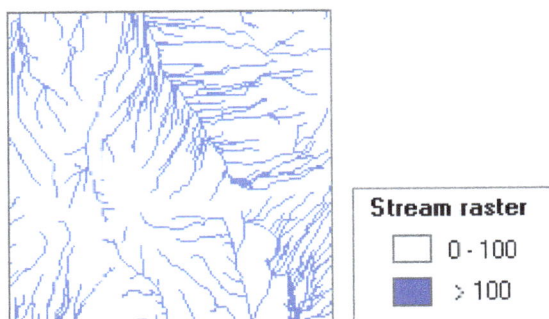

The Stream to Feature tool creates vector stream line features from a stream raster and a flow direction surface.

Network Analysis

A network is referred to as a pure network if only its topology and connectivity are considered. If a network is characterised by its topology and flow characteristics (such as capacity constraints, path choice and link cost functions) it is referred to as a flow network. A transportation network is a flow network representing the movement of people, vehicles or goods.

The approach adopted almost universally is to represent a transportation network by a set of nodes and a set of links. The nodes represent points in space and possibly also in time, and the links tend to correspond to identifiable pieces of transport infrastructure (like a section of road or railway). Links may be either directed, in which case they specify the direction of movement or undirected.

In graph theoretical terminology, a transportation network can be referred to as a valued graph, or alternatively a network. Directed links are referred to as arcs while undirected links as edges. Other useful terms with some intuitive interpretations are a path which is a sequence of distinct nodes connected in one direction by links; a cycle which is a path connected to itself at the ends; and a tree which is a network where every node is visited once and only once. The relationship between the nodes and the arcs, referred to as the network topology, can be specified by a node-arc incidence matrix: a table of binary or ternary variables stating the presence or absence of a relationship between network elements. The node-arc incidence matrix specifies the network topology and is useful for network processing.

Network analysis can be used to solve many different transportation problems that would be very challenging to solve otherwise. A prerequisite to performing network analysis is that you have a network model.

The types of problems which network analysis can be used to solve are quite varied. One common characteristic of the algorithms that power each is that they involve determining the cost of one or more routes through the network. The cost is most commonly based on time or distance, but you can define a cost attribute any way you want. For example, you might score each edge in the network based on its scenic value. You could then create a cost parameter based on the scenic score and use the solver to find the most scenic route.

The Network Data Model

The heart of any GIS is its data model. A data model is an abstract representation of some real-world situation used to organise data in a database. Data models typically consist of three major components. The first is a set of data objects or entity types that form the basic building blocks for the database. The second component is a set of general integrity rules which constrain the occurrences of entities to those which can legally appear in the database. The final component includes operators that can be applied to entities in the database.

Data modelling involves three different levels of abstraction: conceptual, logical and physical levels. Conceptual data models describe the organisation of data at a high level of abstraction, without taking implementation aspects into account. The entity-relationship and the extended entity-relationship models are the most widely used conceptual data models. They provide a series of concepts such as entities, relationships, attributes, capable of describing the data requirements of an application in a manner that is easy to understand and independent of the criteria for managing and organising data on the system. A logical data model translates the conceptual model into a system-specific data scheme, while low-level physical data models provide the details of physical implementation (file organisation and indexes) on a given logical data model.

The network data model is the most popular conceptual model to represent a network within a GIS environment. The model – a special type of the node-arc-area data model that underlies many basic vector GIS databases – is built around two core entities: the Node (a zero-dimensional entity) and the Arc (a one-dimensional entity). Current GIS data models typically represent a network as a collection of arcs with nodes 3 created at the arc intersections. The planar embedding of the node-arc data model guarantees topological consistency of the network.

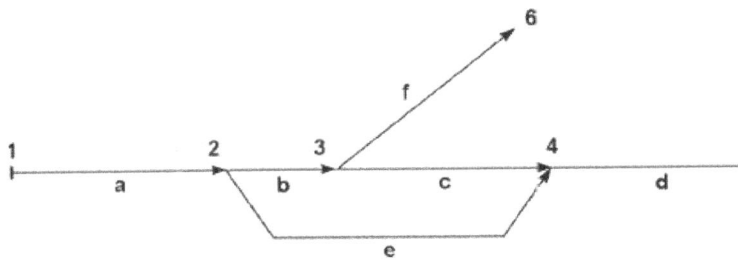

The most widely used logical data model that supports the node-arc representation of networks is the geo relational model. This model separates spatial and attribute data into different data models. A logical spatial data model (the vector data model) that encodes nodes and arcs maintains the geometry and associated topological information, while the associated attribute information is held in relational database management (RDBMS) tables. Unique identifiers associated with each spatial entity (node, arc) provide links to records in the relational model and its data on the entity's attributes. This hybrid data management strategy was developed to take advantage of a relational database management system to store and manipulate attribute information. But this solution does not allow the relationships between a spatial object and its attributes have their own attributes. Though the solution is neither elegant nor robust, it is effective and the geo relational model is widely present in GIS software.

(a) Example network for the relational model example

Arc ID	Street Name	Lanes	Other Attributes
A	High Street	2	
B	High Street	4	
C	High Street	4	
D	High Street	2	
E	River Way	2	
f	Hill Street	2	

(b) A simple arc table

Arc ID	Street Name	Lanes	Other Attributes
1	N	2	
2	Y	4	
3	N	4	
4	Y	2	
5	N	2	
6	N	2	

(c) A Simple node table

Arc ID	Street Name	Lanes	From Node	To Node
A	High Street	2	1	2
B	High Street	4	2	3

C	High Street	4	3	4
D	High Street	2	4	5
E	River Way	2	2	4
F	Hill Street	2	3	6

(d) Pointers added to the arc and node tables to represent connectivity

Node ID	Stop Light?	Arc Links
1	N	A
2	Y	A, b
3	N	B, c, f
4	Y	C, d, e
5	N	D
6	N	f

Figure: Relational data model representations of the arcs and nodes of a network

The relational structure to support the planar network model typically consists of an arc relation and a node relation. The structure may be illustrated as a representation of the simple network shown graphically in figure above. The model implemented in GIS represents each arc of the network as a polyline entity. Associated with each entity will be a set of attributes, conceived as the entries in one row of a rectangular table. Properties may include information about the transverse structure such as the number of lanes or information on address locations within the network. Commonly included attributes are arc length, free flow travel time, base flow and estimated flow. The base and estimated flows usually refer to the observed flow and the flow estimated from some modelling exercise.

The node relation typically contains a node ID field and relevant attributes of the node, such as, for example, the presence of a light. figure (d) shows a scheme that includes storage of pointers from arcs to nodes, and from nodes to arcs, to store the network's topology (connectivity). Each arc has an inherent direction defined by the order of points in its polyline representation. It is noteworthy that the node-arc representation disaggregates a transportation system into separate sub networks for each mode within a single base network. Transfer-arcs link the sub networks. The pseudo-arcs represent modal transfers.

Non-planar Networks and the Turn-table

The planar network data model has received widespread acceptance and use. Despite its popularity, the model has limitations for some areas of transportation analysis, especially where complex network structures are involved. One major problem is caused by the planar embedding requirement. This requirement force nodes to exist at all arc intersections and, thus, ensures topological consistency of the model. But intersections at grade cannot be distinguished from intersections with an overpass or underpass that do not cross at grade. This difficulty in representing underpasses or overpasses may lead to problems when running routing algorithms. The drawbacks in planar topology for network representations have motivated interest in non-planar network models. Planar models force nodes at all intersections, while non-planar network models do not. Non-planar networks are broadly defined as those networks which permit arcs of the network to cross without a network node being located at the intersection. There is no implicit or explicit contact between the line segments at the point of intersections.

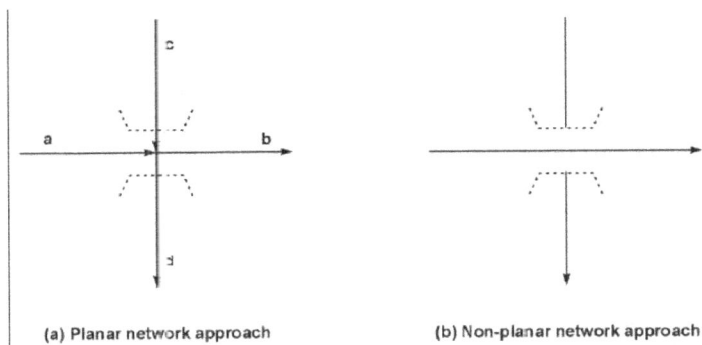

(a) Planar network approach   (b) Non-planar network approach

Figure: Illustration of (a) planar and (b) non-planar network approaches to represent an overpass.

But non-planar data models provide only a partial solution to the problem of connectivity. In transportation network analysis it may be necessary to include extensive information on the ability to connect from one arc to another. Drivers, for example, may force turn restrictions or trucks may be limited by turning radius. Such situations require more than the simple ability to represent the existence of a crossing at grade or an underpass.

To resolve this problem, the standard fully intersected planar network data model has been extended by adding a new structure, called the turn-table. Table shows below a turn-table for the layout used in figure below For each ordered pair of arcs incident at a node, a row of attributes in the table gives appropriate characteristics of the turn (yes/no), together with links to the tables that contain the attributes of the arcs. In this way, a data model with a planar embedding requirement can represent overpasses and underpasses by preventing turns.

From Arc	To Arc	Turn ?
A	C	N
A	B	Y
A	D	N
B	A	Y
B	C	N
B	D	N
C	A	N
C	B	N
C	D	Y
D	A	N
D	B	N
D	C	Y

Table: Layout of a turn-table for the layout used in figure.

Linear Referencing Systems and Dynamic Segmentation

While geographic features are typically located using planar referencing systems, many characteristics associated with a transportation network are located by means of a linear rather than coordinate-based system. These characteristics include data on transportation-related events and facilities (often termed feature data). In order to use linear-referenced attributes in conjunction with a spatially referenced transportation network, there must be some means of linking the two referencing systems together.

Linear referencing systems typically consist of three components a transportation network, a linear referencing method and a datum. The transportation network is represented by the conventional node-arc network. The linear referencing method determines an unknown location within the network using a defined path and an offset distance along that path from some known location. The datum is the set of objects (so-called reference or anchor points) with known geo referenced locations that can be used to anchor the distance calculations for the linear referenced objects.

There are different linear referencing methods. Nyerges identifies three major strategies, namely, road name and kilo meter(mile)point referencing, control section locational referencing, and link and node locational referencing. Road name and kilo-meter point is a system familiar to anyone who has driven on highways in Europe or the USA. This system consists of a road naming convention (that is, a standard procedure for assigning names to highways and streets) and a series of kilo-meter point references (that is, distance calculations along the network, typically measured in fractions of a kilometer or mile). Kilometer point referencing requires a designated point of reference (for example a kilometer 0) as a datum. This is often an end point of the route or where the route crosses a provincial or a national boundary.

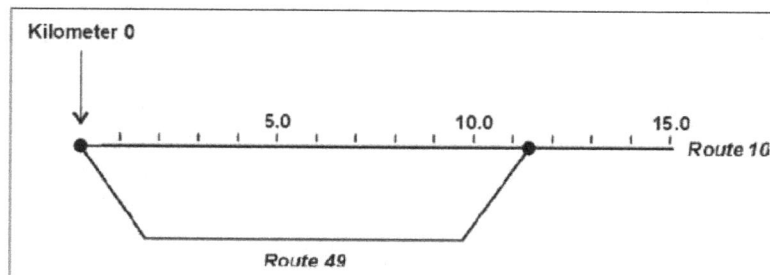

Figure: Kilo-meter point referencing

Due to road modifications and other changes in road geometry, kilo-meter point referencing can become increasingly inaccurate over time. In other words, the reference kilo-meter point may not reflect the actual distance from the point of origin. This may cause problems when maintaining historical records of transportation events. This requires some type of translation factor to adjust distances.

The key to tie (zero-dimensional and one-dimensional) objects located at arbitrary locations on the network to the node-arc structure of the network data model is dynamic segmentation. The term derives from the fact that feature data values are held separately from the actual network route in database tables and then dynamically added to segments of the route each time the user queries the database. Several commercial GIS software packages provide dynamic segmentation capabilities, typically maintained at the logical level using the relational data model. Figure shows a simple illustration of the concept for two types of objects located at arbitrary locations on the network. These entities – termed network points (point events) and network segments (line events) – are given their own attribute tables. Dynamic segmentation reduces the number of transportation features or network links that have to be maintained to represent the system and is particularly useful in situations in which the event data change frequently and need to be stored in a database due to access from other applications.

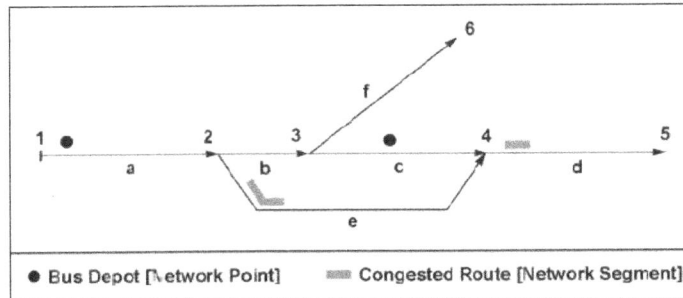

(a)

Arc	Distance from Arc Start	Feature
a	0.7	Bus Depot
c	2.2	Bus Depot

(b)

Arc	Distance from Arc Start to Start of Feature	Distance from Arc Start to End of Feature	Level of Congestion
d	0.5	1.2	High
e	1.0	2.5	High

Figure: The concept of dynamic segmentation: A simple example for (a) network points and (b) network segments

Lanes and Navigable Data Models

A straightforward way to enhance the basic mode-arc model for ITS (intelligent transportation systems) applications is to add information on the transverse structure of the network. Even though certain information about the transverse structure (such as the existence of a median or the number of lanes) might be stored as attributes of arcs or network lines, it is not possible to store detailed information about individual lanes or connectivity at the lane level. There is, for example, no way to disaggregate a turn-table to store turn restrictions which are specific to lanes.

ITS database requirements go well beyond the traditional requirements of maintaining arc-node topology, two-dimensional geo referencing and linear referencing of events within transportation networks. A fully fletched ITS requires a high-integrity, real-time information system which will receive inputs from sensors embedded within transportation facilities and from vehicles equipped with GPS (Global Positioning System) devices and navigable data models. Navigable data models are digital geographic databases of a transportation system that can support vehicle guidance operations of different kinds. For intelligent transportation systems, this includes four functions.

First, the data model has to unambiguously translate coordinate-based locations into street addresses and vice versa. Travellers utilise address systems for location referencing while ITS tracks a vehicle utilising a GPS receiver that can provide locations at accuracies of 5-10 m.

Second, the data model has to support map matching. This refers to the ability to snap a vehicle's position to the nearest location on a network segment when its estimated or measured location is outside the network. This may occur due to differences in accuracy between the digital network database and the global positioning system. Third, the data model has to have the capability to represent the transportation network in detail sufficient to perform different network algorithms, modeling and simulations. In the real world, a transportation network has different types of inter-sections that are of interest to ITS builders. For some applications, information on intersections, lanes and lane changes, highway entrances and exits etc. is important. Other applications may require geometric representation of road curvature and incline. Fourth, the data model must not only assist the traveler in selecting an optimal route based on stated criteria such as travel time, cost and navigational simplicity, but also support route guidance. This refers to navigational in-structions and is a challenging task in real time.

Although dynamic segmentation can be used to enhance the traditional node-arc structure for ITS applications much of the high-resolution positional information provided by in-vehicle GPS receivers is lost when referenced within the traditional network structure. While 50 m accuracy may be sufficient to locate a vehicle on a road, better than 5 m will be required to locate to the lane level. Such accuracies are well beyond the capability of many of the currently available network databases. Achievement of better than 5 m accuracy with GPS requires the use of differential techniques and a high quality of geodetic control.

Fohl describe a prototype lane-based navigable data model where each lane is represented as a distinct entity, with its own connectivity with other lanes, but its geometry is obtained from the standard linear geometry of the road. No attempt is made to store the relative positions of lanes, but the structure does identify such topological properties such as adjacency, and the order of lanes across the road. A more radical approach to navigable data models for IT's is to abandon the node-arc model entirely. Bespalko suggest a 3-D object-oriented GIS-T data model that can distinguish between overpasses, underpasses and intersections and thereby providing guidance through complex intersections.

Route

Network Analyst can find the best way to get from one location to another or to visit several lo-cations. The locations can be specified interactively by placing points on the screen, entering an address, or using points in an existing feature class or feature layer. If you have more than two stops to visit, the best route can be determined for the order of locations as specified by the user. Alternatively, Network Analyst can determine the best sequence to visit the locations, which is known as solving the traveling salesman problem.

Best Route

Whether finding a simple route between two locations or one that visits several locations, people usually try to take the best route. But "best route" can mean different things in different situations.

The best route can be the quickest, shortest, or most scenic route, depending on the impedance chosen. If the impedance is time, then the best route is the quickest route. Hence, the best route can be defined as the route that has the lowest impedance, where the impedance is chosen by the user. Any valid network cost attribute can be used as the impedance when determining the best route.

In the example below, the first case uses time as impedance. The quickest path is shown in blue and has a total length of 4.5 miles, which takes 8 minutes to traverse.

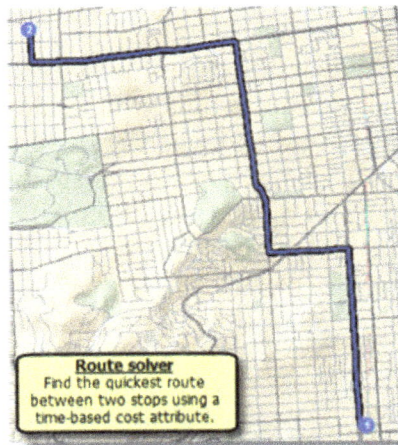

In the next case, distance is chosen as the impedance. Consequently, the length of the shortest path is 4.4 miles, which takes 9 minutes to traverse.

Along with the best route, Network Analyst provides directions with turn-by-turn maps that can be printed.

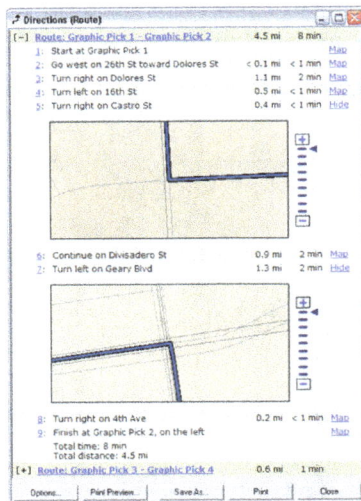

Closest Facility

Finding the closest hospital to an accident, the closest police cars to a crime scene, and the closest store to a customer's address are all examples of closest facility problems. When finding closest facilities, you can specify how many to find and whether the direction of travel is toward or away from them. Once you've found the closest facilities, you can display the best route to or from them, return the travel cost for each route, and display directions to each facility. Additionally, you can specify an impedance cutoff beyond which Network Analyst should not search for a facility. For instance, you can set up a closest facility problem to search for hospitals within 15 minutes' drive time of the site of an accident. Any hospitals that take longer than 15 minutes to reach will not be included in the results.

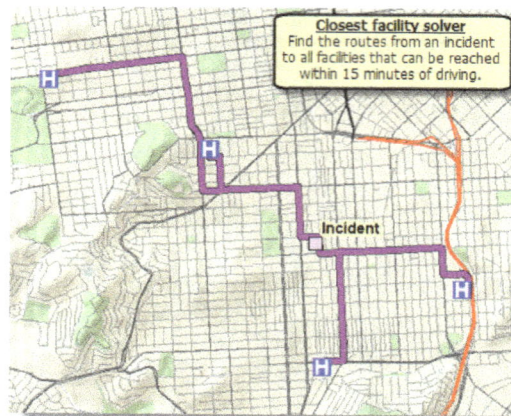

Closest facility solver
Find the routes from an incident to all facilities that can be reached within 15 minutes of driving.

The hospitals are referred to as facilities, and the accident is referred to as an incident. Network Analyst allows you to perform multiple closest facility analyses simultaneously. This means you can have multiple incidents and find the closest facility or facilities to each incident.

Service Areas

With Network Analyst, you can find service areas around any location on a network. A network service area is a region that encompasses all accessible streets, that is, streets that lie within a specified impedance. For instance, the 10-minute service area for a facility includes all the streets that can be reached within 10 minutes from that facility.

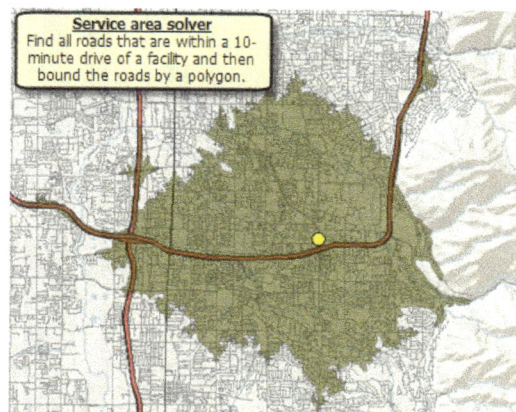

Service area solver
Find all roads that are within a 10-minute drive of a facility and then bound the roads by a polygon.

Accessibility

Accessibility refers to how easy it is to go to a site. In Network Analyst, accessibility can be measured in terms of travel time, distance, or any other impedance on the network. Evaluating accessibility helps answer basic questions, such as How many people live within a 10-minute drive from a movie theatre, How many customers live within a half-kilometer walking distance from a convenience store? Examining accessibility can help you determine how suitable a site is for a new business. It can also help you identify what is near an existing business to help you make other marketing decisions.

Evaluating Accessibility

One simple way to evaluate accessibility is by a buffer distance around a point. For example, find out how many customers live within a 5-kilometer radius of a site using a simple circle. However, considering people travel by road, this method won't reflect the actual accessibility to the site. Service networks computed by Network Analyst can overcome this limitation by identifying the accessible streets within five kilometers of a site via the road network. Once created, you can use service networks to see what is alongside the accessible streets, for example, find competing businesses within a 5-minute drive.

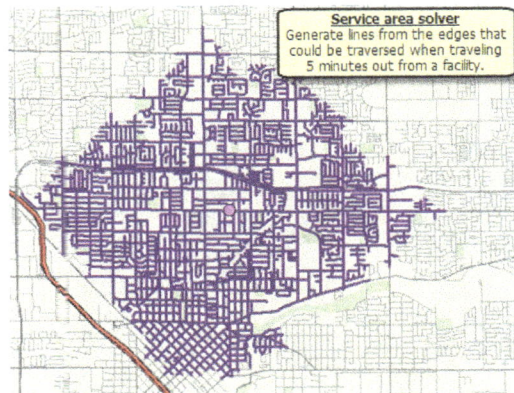

Multiple concentric service areas show how accessibility changes with an increase in impedance. It can be used, for example, to show how many hospitals are within 5-, 10-, and 15-minute drive times of schools.

By solving with traffic data, you can see which hospitals can be reached within these drive times for different times of the day. The reachable hospitals may change due to traffic conditions.

OD Cost Matrix

With Network Analyst, you can create an origin-destination (OD) cost matrix from multiple origins to multiple destinations. An OD cost matrix is a table that contains the network impedance from each origin to each destination. Additionally, it ranks the destinations that each origin connects to in ascending order based on the minimum network impedance required to travel from that origin to each destination.

The best network path is discovered for each origin-destination pair, and the cost is stored in the attribute table of the output lines. Even though the lines are straight for performance reasons, they always store the network cost, not straight-line distance. The graphic below shows the results of an OD cost matrix analysis that was set to find the cost to reach the four closest destinations from each origin.

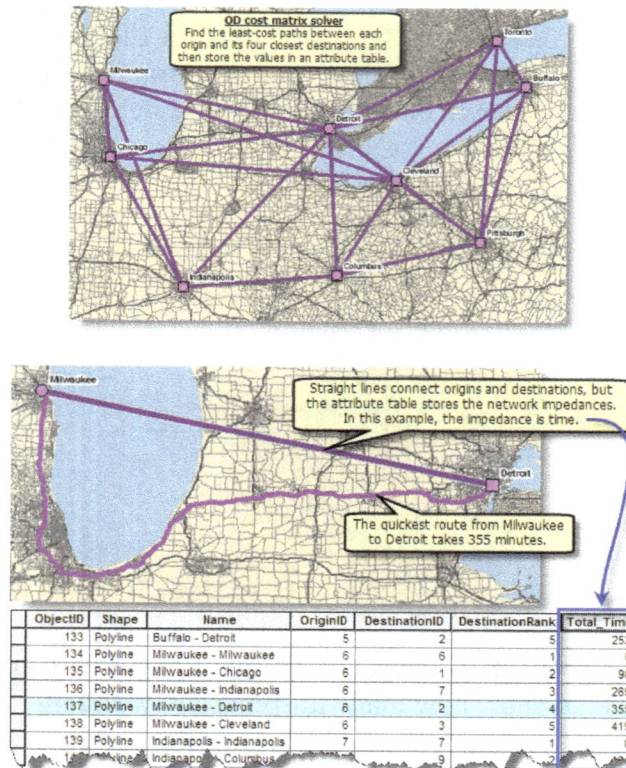

ObjectID	Shape	Name	OriginID	DestinationID	DestinationRank	Total_Time
133	Polyline	Buffalo - Detroit	5	2	5	252
134	Polyline	Milwaukee - Milwaukee	6	6	1	0
135	Polyline	Milwaukee - Chicago	6	1	2	98
136	Polyline	Milwaukee - Indianapolis	6	7	3	285
137	Polyline	Milwaukee - Detroit	6	2	4	355
138	Polyline	Milwaukee - Cleveland	6	3	5	419
139	Polyline	Indianapolis - Indianapolis	7	7	1	0

The straight lines can be symbolized in various ways, such as by color, representing which point they originate from, or by thickness, representing the travel time of each path.

Vehicle Routing Problem

A dispatcher managing a fleet of vehicles is often required to make decisions about vehicle routing. One such decision involves how to best assign a group of customers to a fleet of vehicles and to sequence and schedule their visits. The objectives in solving such vehicle routing problems (VRP) are to provide a high level of customer service by honoring any time windows while keeping the overall operating and investment costs for each route as low as possible. The constraints are to complete the routes with available resources and within the time limits imposed by driver work shifts, driving speeds, and customer commitments.

Network Analyst provides a vehicle routing problem solver that can be used to determine solutions for such complex fleet management tasks.

Consider an example of delivering goods to grocery stores from a central warehouse location. A fleet of three trucks is available at the warehouse. The warehouse operates only within a certain time window—from 8:00 a.m. to 5:00 p.m.—during which all trucks must return back to the warehouse. Each truck has a capacity of 15,000 pounds, which limits the amount of goods it can carry. Each store has a demand for a specific amount of goods (in pounds) that needs to be delivered, and each store has time windows that confine when deliveries should be made. Furthermore, the driver can work only eight hours per day, requires a break for lunch, and is paid for the amount spent on driving and servicing the stores. The goal is to come up with an itinerary for each driver (or route) such that the deliveries can be made while honoring all the service requirements and minimizing the total time spent on a particular route by the driver. The figure below shows three routes obtained by solving the above vehicle routing problem.

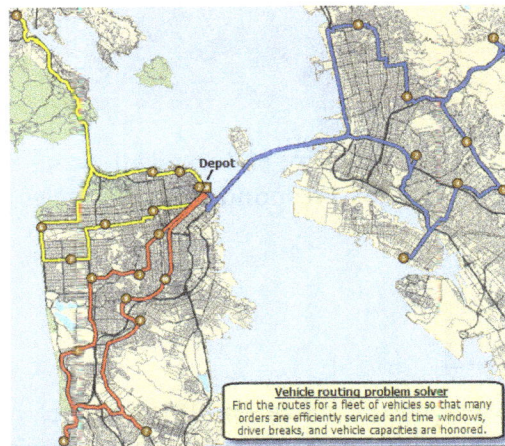

Vehicle routing problem solver
Find the routes for a fleet of vehicles so that many orders are efficiently serviced and time windows, driver breaks, and vehicle capacities are honored.

Location-allocation

Location-allocation helps you choose which facilities from a set of facilities to operate based on their potential interaction with demand points. It can help you answer questions like the following:

- Given a set of existing fire stations, which site for a new fire station would provide the best response times for the community?

- If a retail company has to downsize, which stores should it close to maintain the most overall demand?

- Where, should factory be built to minimize the distance to distribution centers?

In these examples, facilities would represent the fire stations, retail stores, and factories; demand points would represent buildings, customers, and distribution centers.

The objective may be to minimize the overall distance between demand points and facilities, maximize the number of demand points covered within a certain distance of facilities, maximize an apportioned amount of demand that decays with increasing distance from a facility or maximize the amount of demand captured in an environment of friendly and competing facilities.

The map below shows the results of a location-allocation analysis meant to determine which fire stations are redundant. The following information was provided to the solver: an array of fire stations (facilities), street midpoints (demand points), and a maximum allowable response time. The response time is the time it takes firefighters to reach a given location. The location-allocation solver determined that the fire department can close several fire stations and still maintain a three-minute response time.

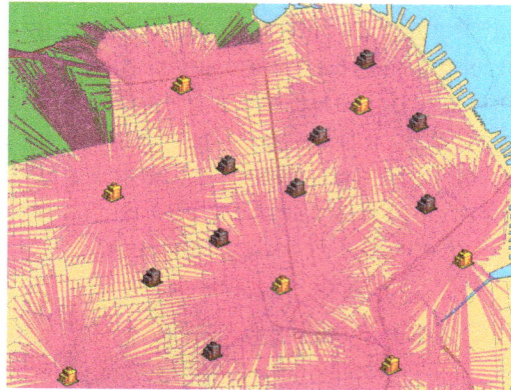

Out of the current set of fire stations, nine fire stations can close, and a minimum of seven are needed for the department to still be able to respond to emergencies within three minutes.

Geovisualization

Geovisualization is short for geographic visualization. It is a branch or discipline within visualization that deals solely with displaying information that has a geospatial component to it. A geospatial component is geographic or positioning information. For example, consider the weather report on the evening news. You typically have precipitation information of some sort (fog, rain, or snow) displayed over a map of the local area as a background. The visual gives you a detailed look at how much precipitation you will get, where in the local area it will appear, and at what time of day. This is Geovisualization hard at work converting raw weather and positioning information into a graphic form that we can understand in a matter of seconds.

Geo-visualization is a rather new tool for understanding, interpretation and assessment of environmental data. Geo-visualisation of landscape patterns and processes offers an innovative approach to present and discuss results of data collection, data analysis and data simulation. Cartography and scientific graphics form the roots of geo-visualisation however it is all based on the digital principles of Geographical Information Systems.

Hence, geovisualisztion represents a powerful tool that can be of strategic importance for all disciplines in environmental sciences. It offers new perspectives in research and practice, especially with respect to public and stakeholder involvement making research results and predictions visible to laymen. Geo-visualization does not only offer the possibility to show developments in the past, but also to visualize possible future scenarios, depending on alternative decisions that can be taken today.

Goals of Geovisualization

The goals of geovisualization are manifold. The map use cube by MacEachren and Kraak models the space of visualization goals with respect to three dimensions:

- The task can range from revealing unknowns and constructing new knowledge to sharing existing knowledge;

- The interaction with the visualization interface can range from a rather passive low level to a high level where users actively influence what they see;

- Finally, the visualization use ranges from a single, private user to a large, public audience.

The four visualization goals exploration, analysis, synthesis, and presentation are placed on a diagonal in this map-use space. On the one extreme, exploration can be found as a private, highly interactive task to prompt thinking and to generate hypotheses and ultimately new scientific insight. The other extreme is formed by presenting knowledge in low-interaction visualizations to a wide audience, e.g., on a professional conference or in a publication described these two extremes as visual thinking which creates and interprets graphic representations, and visual communication which aims at distributing knowledge in an easy-to-read graphic form. The former task is exploratory, while the latter one is explanatory.

In the beginning of the 1990s, geovisualization research focused on exploratory methods and tools. Communication was the realm of traditional cartography. Every map communicates a message by stressing certain aspects of the underlying data. Cartographers, due to the lack of interaction in paper maps, had the goal of finding an optimal map for the intended message. In exploration, where the 'message' is yet to be discovered there is no optimal map in the beginning. The early focus on exploration has expanded recently to include the whole range of visualization tasks as Dykes observed. The reason is that sophisticated interactive geovisualization methods are now recognized as useful not only for exploration but also for presentation of knowledge through guided discovery. The visualization experience offers great benefits for understanding and learning as it enables both experienced scientists and students to (re)discover knowledge through interaction.

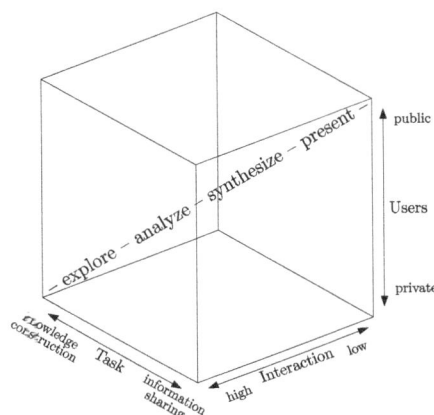

Figure: The map use cube after MacEachren and Kraak characterizing geovisualization objectives in a three-dimensional space by their level of interaction, their audience, and the addressed tasks.

The map is now frequently seen as an interactive interface to access and explore geospatial data while it still retains its traditional role as a presentational device. Dykes argued that interaction appears to be the key defining characteristic of geovisualization today and MacEachren and Kraak stated that geovisualization is characterized by interaction and dynamics. Concerning previously private tasks such as exploration, a shift from individual use towards support of group work has been demanded in the ICA agenda. So recently, in terms of the map use cube, more research efforts have been attracted by the high-interaction and group-use (or public) parts of the geovisualization space.

Driving Forces of Geovisualization

The first is the rapid advances that have been made in graphics and display technology. The availability of both low-cost 3D graphics hardware in personal computers and the development of highly immersive 3D virtual environments resulted in investigating the potential that these technologies have for visualizing geospatial data. However, this emphasis on realism contrasts with the history of cartography that point to centuries of successful abstraction making the world easier to understand according to MacEachren. Indeed, maps filter out unnecessary details of the environment in order to highlight interesting information. For example, a road map based on satellite images would be extremely hard to use. The challenge is to study the relative advantages and disadvantages of realism and abstraction in geovisualization and then, depending on the problem context, potentially integrate both abstract and realistic displays in a single geovisualization environment.

The second driving force for geovisualization is the need to analyze and explore a dramatically increasing amount of geospatial data that are routinely collected these days by a multitude of scientific and governmental institutions, private companies, and individuals. This is due to an increasing availability and decreasing cost of technology to acquire, store, and process these data. For example, the location of a credit card purchase or a mobile phone call is recorded by computers. A majority of the data, MacEachren and Kraak estimated up to 80 percent, contain geospatial references, e.g., coordinates of environmental measurements, census data, positions of vehicles, ships, planes, and parcels, addresses of customers, etc. These data, often characterized by a high dimensionality, are a vast source of potentially valuable information for research and decision making, e.g., in studying disease incidence patterns, traffic flows, credit card fraud, or climate change. Privacy issues with these data are an important concern but they are out of the scope of this chapter. The large volume of many data sets poses challenging problems for their exploration. While computers are well suited for processing large amounts of data or finding well-known patterns, they perform rather poorly in detecting and interpreting unknown patterns in noisy data—at least in comparison to the human brain. On the other hand, with increasing data volume and complexity humans quickly reach the limit of their capacities in analyzing raw numeric and textual data. The goal of geovisualization is to combine the strengths of human vision, creativity, and general knowledge with the storage capacity and the computational power of modern computers in order to explore large geospatial data sets. One way of doing this is by presenting a multitude of graphic representations of the data to the user, which allows him or her to interact with the data and change the views in order to gain insight and to draw conclusions.

Finally, the third driving force for geovisualization is the rise of the Internet and its development into the prominent medium to disseminate geospatial data and maps. On the one hand, the Internet facilitates collaboration of expert users at different places, which is one of the ICA Commission's research challenges, and, on the other hand, it enables geovisualization applications to address the public. Reaching the public is an important aspect both for governmental agencies and for business companies who provide and sell services based on geospatial information.

Cognitive Aspects

Visual thinking describes mental information processing through images instead of words. DiBiase saw the origins of the potential power of visual thinking in the biological evolution where individuals with a quick reaction to visual cues survived. While we communicate mostly through words, we are connected to our environment primarily through vision. Hence, our visual perception has evolved into a powerful system that actively seeks meaningful patterns in what we see, sometimes even imposing them where they do not exist. Visual thinking often does not follow a logical train of thought and hence has not been appreciated for a long time in science. However, in 1990, MacEachren and Ganter reported a renewal of interest in human vision as a tool of advancing science.

Prominent examples of successful visual thinking are Wegener's theory of continental drift prompted by the similar shapes of the facing coasts of Africa and South America, Kekul´e's ring-shaped model of the benzene structure, or the discovery of the double helix structure of DNA by Watson and Crick stimulated by an x-ray photograph of DNA.

MacEachren and Ganter developed a cognitive approach to geovisualization. They concentrated on the explorative side of visualization and thus on gaining new scientific insight through visual thinking. In their article visualization is seen in the first instance as a mental process rather than generating images using computer graphics. Nonetheless, computer graphics are a valuable tool to stimulate this mental process by creating graphics that facilitate visual identification of patterns or anomalies in the data. Visualization should not primarily focus on generating images but on using images to generate new ideas. This also means that elaborate and highly realistic images are not necessarily required to generate valid hypotheses. Instead it is often abstraction, in the past achieved with pencil and paper that helps to distinguish pattern from noise and thus makes a map or some other graphic useful. One key aspect in visual data exploration is to view a data set in a number of alternative ways to prompt both hypotheses and their critical reflection.

Figure: John Snow's map of cholera deaths in London 1854.
Deaths are marked by dots and water pumps by crosses.

An early example of how a cartographic picture was used to gain new insight comes from medicine. In 1854 the London physician John Snow mapped cholera cases to a district map and made the link between cholera and a specific water pump that was used by the infected persons who 'clustered' around that pump on the map. In fact, it was reported that an anomaly of that pattern finally prompted his insight, namely the case of a workhouse with very few infections in the center of the cholera outbreak: it had an independent water source.

In order to successfully facilitate visual thinking it is necessary to understand how the human mind processes visual information. MacEachren, Ganter and MacEachren described visual information processing. Essentially, human vision produces abstractions from the complex input on the retina and these abstractions are matched to the mind's vast collection of patterns (or schemata) from experience.

MacEachren and Ganter proposed a two-stage model for interacting with geovisualization tools in scientific exploration. At the first stage, called seeing-that, the analyst searches for patterns in the visual input. They distinguished two types of pattern-matches: a pattern is recognized if it is expected in the context; noticing, however, means detecting unexpected patterns that might lead to new insight. Once a pattern is recognized or noticed the analyst enters the second stage called reasoning-why, also known as the confirmatory stage of scientific inquiry. At this stage, the judgment made before is carefully examined to identify errors or to explain a pattern or anomaly. These two steps are iterated to collect more evidence for or to modify a judgment.

Finally, when the scientist has confirmed a hypothesis, he or she will usually want to share his insight with scientific peers through presentations and publications. Now, the goal is to lead fellow scientists to the same insight by invoking their seeing-that and reasoning-why process through well-designed graphics. If fellow scientists discover the patterns themselves the author's arguments will be much more convincing.

The model of MacEachren and Ganter implies that the success of a geovisualization tool in scientific inquiry depends primarily on its ability of displaying patterns that can be identified and analyzed by a human viewer. However, individual users recognize and notice patterns differently based on their individual experience. Hence, explorative geovisualization tools must be interactive and permit a wide range of modifications to the visual display of the data. In the reasoning-why process, a key to insight or error detection is examining a judgment from different perspectives. Errors in pattern identification are divided into two categories: Type I errors mean seeing-wrong, i.e., to see patterns where they do not exist. Type II errors, on the contrary, denote not-seeing, i.e., missing patterns that are really there. Since human perception is adapted to seeing patterns all the time humans are susceptible to Type I errors (e.g., seeing shapes in clouds). Conversely, there is an effect known as scientific blindness, a phenomenon describing the tendency to overlook what one is not actively searching for. Consistency of patterns across multiple perspectives and modes is a cue to a valid pattern while inconsistencies demand reconsidering or rejecting the pattern.

Graphic Variables

On a map, information is usually represented by symbols, points, lines, and areas with different properties such as color, shape, etc. Bertin's concept of fundamental graphic variables for map and graphic design and rules for their use, published as "S´emiologie graphique" in 1967, has proposed a basic typology for map design. This work was based on his experience as geographer and cartographer. Since then his original set of variables has been modified and extended, graphic variables, size, density/size of texture elements, color hue, color saturation, color value, orientation, and shape, are means of communicating data to a map reader. Especially the different variables of color have been studied with regard to their efficiency in representing different kinds of data (categorical, sequential, binary, etc.). Variables can also be combined to represent the same information redundantly, for example using both size and color value. This provides better selectivity and facilitates the judgment of quantitative differences on a map in comparison to using just one variable.

Originally, Bertin's variables have been designed to describe information visualization on paper maps. Today, advances in graphics display technology provide a set of new graphic variables that can be utilized in geovisualization. Transparency and crispness are regarded as static graphic variables the latter for example is suitable to represent uncertainty of some classification on a map. However, geovisualization goes beyond static maps and therefore sets of tactile, dynamic, and sonic variables have been proposed, e.g., loudness, pitch, duration, temporal position, rate-of-change, etc. Most of these variables are analogs of graphic variables in another dimension, e.g., duration corresponds to size and temporal position to spatial location. However, both dynamic and sonic variables need to be observed over time and thus require more user attention than static representations.

Visualization Methods and Techniques

Geospatial Data

In contrast to information visualization displaying any abstract data, geovisualization deals specifically with geospatial data, i.e., data that contain georeferencing. This is a unique feature and special methods and tools are needed to take full advantage of it. Haining decomposed geospatial data into an abstract attribute space and a geographic space with two or three dimensions. According to MacEachren and Kraak geospatial data and information are fundamentally different from other kinds of data since they are inherently spatially structured in two or three dimensions (latitude, longitude, and possibly altitude). In case of spatiotemporal data, time can be seen as a fourth dimension. Remaining dimensions are often unstructured and there are various visualization methods for these abstract data. Note also that distances and directions have an immediate meaning in those dimensions in contrast to distances computed on abstract data. While general visualization methods may be applied to spatial data as well, they do not take into account the special characteristics of the attributes. The georeference is usually either to a single point or to a whole area. Whether geospatial data are defined as a point or as an area obviously depends on the geographic scale at which they are examined. For example, a village can be represented as an area on a large scale map, as a point on a map of its province, and not at all on a country level.

2D Cartographic Visualization

The most common visualization method for geospatial data is a cartographic display of some form, i.e., a map where the area under consideration is depicted, and onto which the data of interest are plotted at their corresponding coordinates. Space is used to depict space by mapping latitude and longitude to the coordinate axes of the map drawing area. This might seem to be the most natural way of using this graphic variable. However, there are good reasons of linking for example population to space resulting in a cartogram1 where area on the map is not proportional to a certain geographic area but in this case to the number of people living in that very area. Still, cartograms usually try to preserve the users' mental map by keeping similar shapes and by preserving the adjacencies between the depicted areas. An example of a world population cartogram is shown in figure below Tobler gave an overview of algorithms to automatically create contiguous value-by-area cartograms and van Kreveld and Speckmann studied drawing rectangular cartograms. Especially when the focus of a map is on social, economic, or political issues, cartograms help to draw the users' attention to population as the map's theme while avoiding to emphasize large but sparsely inhabited regions. MacEachren described space as an indispensable graphic variable with a large influence on what the user of a map sees. He argued that therefore space should represent the map theme.

For both cartograms and geographic maps the interesting aspect is of course how to depict abstract attributes of the data or at least a subset of them. Among the most popular methods to represent categorical but also numerical data are choropleth2 maps. A choropleth map uses the graphic variables describing properties of color or texture to show properties of non-overlapping areas, such as provinces, districts, or other units of territory division. A number of categories is mapped to distinct colors or textures which are used to fill the areas accordingly. Examples are land cover/use with categories like forest, crop land, housing, etc. or election results per district, e.g., displaying the percentage of votes for a certain party or the voter turnout as in figure above for unordered data well-distinguishable colors are needed while for ordered data it is important to find a lightness or hue scale that represents the original range of numbers efficiently, i.e., that the user can estimate values and differences from the colors.

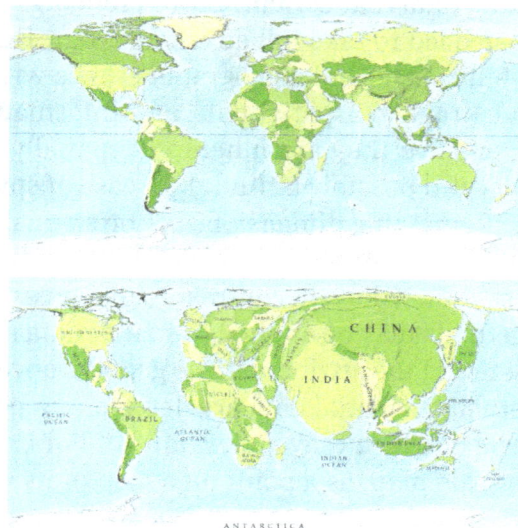

Figure: Geographical world map (top) and world population cartogram (bottom).
In the cartogram the size of a country is proportional to the size of its population.

Alternatively, a continuous range of attribute values is mapped to a continuous color range without assigning values to a fixed number of classes. While choropleth maps help to show general trends in the data there is certainly a loss of information because the user cannot map a certain color to its exact numerical value. Furthermore, a choropleth map can only express one or two attributes of the data (by using a two-dimensional color scheme or by combining color and texture). Andrienko described a selection of methods to represent single and multiple attributes in a map. Depending on the type of the attributes (logical, numeric, or nominal), they used bar and pie diagrams common in statistic visualization. Similarly, glyph-based techniques from visual data mining can also be combined with map displays. These techniques, described in more detail later, use compound glyphs to represent the values of multiple abstract attributes. Using their geospatial reference, glyphs or statically diagrams are placed on the map and thus both spatial and multidimensional abstract attributes are represented on a single map.

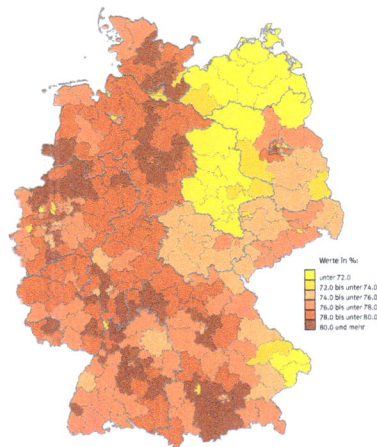

Figure: A choropleth map showing turnout of voters
in the 2005 federal elections in Germany.

However, if the number of symbols or attributes exceeds a certain limit the symbols become hard to compare and other non-map based techniques from visual data mining should be applied in addition to the display of a map or cartogram.

Other approaches for displaying high-dimensional data reduce the dimensionality of the data, e.g., by applying statistical techniques like principal component analysis or by calculating compound indices representing, for example, the socioeconomic development of a region. The disadvantage, especially for explorative visualization, is that through the loss of information potential patterns of some attributes might get lost.

3D Cartographic Visualization

In contrast to traditional paper maps and two-dimensional visualization methods, geovisualization can go one step further and use the potential of increasingly experiential representation technologies. 3D visualization includes the full range from regular 3D graphics hardware in desktop computers to immersive 3D displays, CAVEs (Cave Automatic Virtual Environments), and Power Walls providing stereoscopic views. Since, humans live in a three-dimensional environment our perception and cognition is adapted to processing 3D visual stimuli. But there is still little known about when 3D visualization is appropriate and how it can effectively enhance visual thinking.

Cartography has a long and successful tradition using abstraction to depict a wide range of data on maps. In contrast, the focus of computer graphics technology is on producing increasingly realistic images and virtual environments. Virtual reality techniques are widespread, for example, in architecture and landscape planning where realism is very important. Depending on the geovisualization task, realism can be a distraction and insight is more likely when using abstract visual symbolism. But as MacEachren et al. pointed out, there had been only few efforts exploring abstract visualizations of geospatial data in 3D.

In terms of the 3D representation of the data, MacEachren et al. distinguished between using the three dimensions of the representation to display the three dimensions of physical space, using one or two dimensions for non-spatial data, e.g., income or time, and using all three dimensions for abstract data. Representing time as the third dimension is common on spatiotemporal visualization.

Today, the most widespread use of 3D is at the level of visual representation while the display is a 2D screen. It is important to be aware of the implications that the projection of a 3D representation onto a 2D plane has. Depth cues such as perspective and occlusion also cause problems because distances are harder to estimate, and occlusion hides objects depending on the viewpoint. Ware and Plumlee observed that due to occlusion humans cannot perceive much information in the depth direction while the x- and y-directions, orthogonal to the line of sight, can convey complex patterns. A set of interactive navigational controls are necessary to move within the 3D representation, e.g., zooming or flying, cf. As Wood et al. pointed out the effectiveness of the virtual environment metaphor relies to some extent on navigational realism. While moving (e.g., walking or flying) slowly through the visual space maintains a sense of orientation, faster modes of movement such as teleporting lose the context and the user has to reorient ate.

Visual Data Mining Tools

Visual data-mining, also denoted as exploratory data analysis (EDA) in statistics by Tukey, is a human centre task that aims at visually analyzing data and gaining new insights. This contrasts computational data mining techniques which use algorithms to detect patterns in the data. Effective visual data mining tools need to display multivariate data in a way that the human viewer can easily perceive patterns and relationships in the data. Visual data mining in general is not tailored specifically for geospatial data. for a Since geospatial data usually have many abstract attributes these general techniques can be applied for displaying non-spatial attributes of the data. Visualization techniques for multivariate data were broadly classified as geometric, glyph- or icon-based, pixel-oriented, and hierarchical by Schroeder and Keim and Kriegel. In a geovisualization context, geometric and glyph-based techniques are most common.

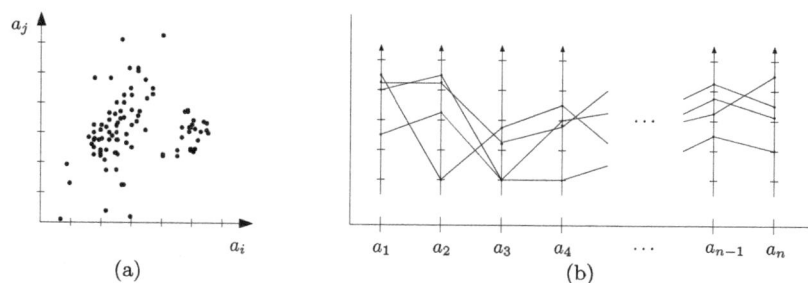

Figure: Example of a 2D scatter plot in subfigure
(a) and a parallel coordinate plot in subfigure (b).

Geometric Techniques

Geometric Techniques Two geometric techniques commonly used in geovisualization are scatter plots and parallel coordinate plots. Scatter plots in their basic two-dimensional form depict objects as points in a coordinate system where the axes correspond to two selected attributes, Elements in the data set with similar values in these attributes form visual clusters in the scatter plot. The idea of a scatter plot can be extended to three dimensions but then phenomena as occlusion and different perception of depth. The extension to more than three dimensions is often implemented by drawing a scatter-plot matrix containing one scatter plot for each pair of attributes. This, however, makes the identification of multidimensional patterns difficult because many plots in the matrix need to be linked mentally.

Parallel coordinate plots (PCP) are a means of displaying high dimensional data in a single plot. In a PCP, one dimension is used to place multiple parallel axes, each of which represents one attribute of the data. Each element of the data set is then characterized by the values of its attributes which are connected along the axes and thus build a geometric profile of that element. Since all elements are plotted in this way, the user can identify similar objects by comparing the geometric shape of their profiles. However, depending on the number of profiles, over plotting occurs and may result in poor legibility. Keim and Kriegel estimated that about 1,000 items could be displayed at the same time. Moreover it becomes difficult to compare profiles based on an increasing number of attribute axes. Another important aspect of PCPs is the order of the attributes plotted along the parallel axes since this order has a strong influence on the shapes of the profiles. Hence, a user should be able to rearrange the attributes manually or based on sorting algorithms.

Glyph-based Techniques

Glyph-Based Techniques Glyph-based or icon-based techniques use a mapping of multiple attribute values to a set of different visual features of a glyph which in turn represents one data object. Two examples of such techniques are Chernoff faces and star plots. In a Chernoff face, different variables of the data are related to facial features of an iconic face, such as size and shape of mouth, eyes, ears, etc. The motivation of using faces to depict multidimensional data is that human mind is used to recognize and compare faces. However, different features, e.g., shape of the eyes and area of the face, are hard to compare and Chernoff faces are, in contrast to human faces, not perceived pre-attentively such that there is no advantage over other types of glyphs.

Star plots depict the value of a set of attributes by the length of rays emanating from the center of the glyph. While the maximum number of facial features in Chernoff faces is reached quickly, star plots can display data with higher dimension by increasing the number of rays. Again, as for parallel coordinate plots, the order of the attributes influences the shape of the star plots.

A nice property of glyph-based techniques is that they can be easily combined with map displays by placing each glyph according to its geospatial coordinates on the map. However, with an increasing number of symbols or attributes glyph-based techniques are of limited use due to the difficulty of visually recognizing patterns and distinguishing features on a display with too many glyphs or glyphs with too many features.

Graph-drawing Techniques

Graph-Drawing Techniques Geospatial data often contain links between related elements, e.g., routes, trade connections, etc. Exploring such data sets includes the search for patterns in the link structure between items. Data containing relationships between elements are mathematically modeled as a graph consisting of a set of nodes, the data elements, and a set of (weighted) edges, the links between elements. The research area of graph drawing provides a multitude of algorithms for visualizing such graphs.

In general, for graph drawing the emphasis is on finding a layout, i.e., positions of nodes and edges of a given graph that satisfies certain aesthetic criteria, e.g., few edge crossings. In geovisualization, there are usually certain constraints on such a layout since nodes already have a spatial location. In that case, finding a legible layout for the edges is of interest, for example in schematizing road networks. In other cases, such as drawing metro maps, the network topology is more important and node positions are only required to satisfy certain relative positions (e.g., left, right, above, below) in order to preserve the user's mental map. Finally, some data is best analyzed by putting no restrictions to node positions and using a general algorithm to find a graph layout in which link patterns can be identified visually. Such methods are applied in visual social network analysis, for example Brandes and Wagner. In the latter cases, where node positions are modified, a map display of the true geography in combination with a graph layout focusing on the link topology is helpful for identifying both spatial and link-based patterns.

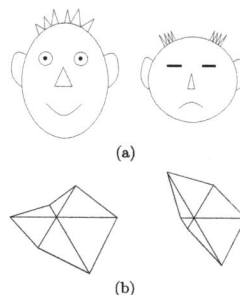

Figure: Two examples of glyphbased techniques: subfigure (a) shows two Chernoff faces and subfigure (b) shows two star plots on six attributes.

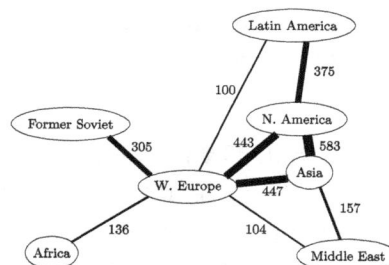

Figure: A graph showing trade relationships. Edges are weighted by trade volume and drawn shorter and thicker with increasing weight. Only edges with at least 100 billion dollars trade volume are shown.

An example by Rodgers visualizing trade volume between regions of the world as a graph is shown in figure. The stronger a trade relationship between two regions the more they attract each other in the graph layout.

Spatio-temporal Visualization

Spatio-temporal data are very common in the earth sciences and related disciplines. The goal of many studies is to reveal, analyze, and understand patterns of temporal change of phenomena, such as global warming, population development, or spread of diseases. Animation works well in displaying patterns if they are based on the same temporal sequence as the animation itself, e.g., showing trends like urban growth over time. However, Andrienko et al. criticized that for less evident patterns it is necessary to compare the data at different points in time which involves memorizing a large number of states in an animated display, even if interactive controls allow to pause, jump, and step through specific points in time. Thus it might be more effective to statically display selected moments in time simultaneously using small multiples. Then, an analyst can directly compare attribute properties of different points in time at his or her own speed. However, the number of simultaneous images on the screen is limited and long time series have to be evaluated piecewise. Andrienko et al. argue that, for all these reasons, spatio-temporal data exploration must be supported by a variety of techniques, possibly in combination with an animated display.

Andrienko et al. classified spatio-temporal data according to the type of temporal changes:

1) Existential changes, i.e., appearance and disappearance of features.

2) Changes of spatial properties, i.e., change of location, shape, size, etc.

3) Changes of thematic properties, i.e., qualitative and quantitative changes of attributes.

Following this classification, they presented corresponding visualization techniques. All techniques involved a map display to visualize the spatial attributes of the data.

Data of existential changes usually consist of events or observations at specific moments or time periods during which a certain property holds, e.g., road congestion data. Hence a map showing these data always considers a selected time interval. If data items are represented by glyphs, one way to display the time associated with them is by using textual labels. Another possibility is using a color scheme to represent the age of the data. A 3D representation of space and time is a third and common method. In such a space-time cube, the third dimension corresponds to time while two dimensions represent geographical space. The reference map is usually displayed in the coordinate plane corresponding to time 0 and data items are positioned above the map depending on their spatial locations and their times of appearance.

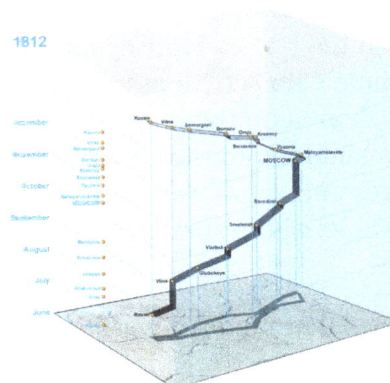

Fig: A space-time cube visualization

An example of a space-time cube is for data that contain moving objects, comparing object trajectories is of interest. Static 2D maps are able to show the trajectories of a small number of objects but in this simple form it is not possible to evaluate aspects like speed or whether two objects met at a crossing or just visited at different points in time. Andrienko et al. suggested animating object movements, either as a sequence of snapshots in time in which at each moment objects are shown at their current positions or using the movement history and showing the trajectories up to the current point in time. Movement history can optionally be limited to a specified time interval. It was found that the snapshot technique was suited for a single object while several objects were better observed displaying also the movement history. MacEachren suggested using the space-time cube to display trajectories which avoids the disadvantages of 2D trajectories mentioned above as it shows when and not just if an object visited a point.

There are several methods of displaying thematic attributes on a map. A very effective and common method is the choropleth map. Animating a choropleth map is able to give a good overview of the values in a selected attribute. However, it is difficult to estimate trends in a particular area on the map or to compare trends between different areas.

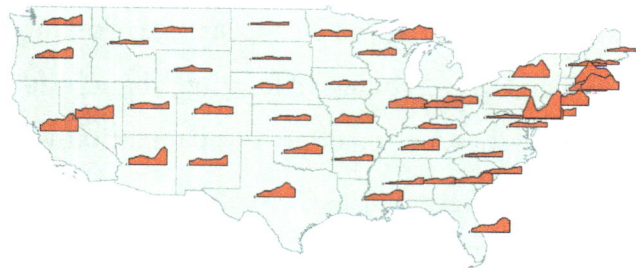

Figure: Cartographic representation of the spatial
distribution of the burglary rates in the USA.

Change maps, adapted from conventional cartography, use the choropleth map to show the differences of an attribute between two selected points in time. Mapping increase and decrease in attribute value to shades of two different color allows to evaluate regional changes for two moments. Such a map is restricted to two points in time and the map can be misleading because information on the actual attribute values is lost, e.g., concerning crime data two areas can have very different burglary rates but still be colored the same if the rates both decrease by the same value. Andrienko combined time-series graphs with maps to avoid these disadvantages. A time-series graph is a two-dimensional plot of the temporal variation of attribute values, where, time is represented on the x-axis and attribute values on the y-axis. Plotting all data in the same time graph gives an overview of the dynamics of the whole data set. To assess local behaviors, Andrienko plotted the time-series data individually for each area on the map and used the closed shape of the plot as a symbol superimposed on each area similarly to the glyph-based techniques for an example. This technique allows to evaluate changes and actual values of an attribute for the whole time period under consideration. The user can explore both spatial patterns and patterns in the attribute space in the same view.

Shanbhag presented three techniques that modify choropleth maps in order to display temporal attribute data. They did not color each district area in the map uniformly but partitioned it into several regions, each representing one point of time in the data. Their first technique builds on a cyclical, clocklike metaphor for perceiving time and partitions the area polygon into wedges. The second technique draws from the metaphor of annual rings of a tree trunk and assigns time points to 'rings' of the polygon. Finally, they suggested time slices for a linear perception of time, i.e., polygons were partitioned into vertical slices. Using any of the three techniques, temporal trends in each area could be detected by observing the variation (e.g., brightness) of the different regions of the district area. However, for an effective visualization the number of simultaneously displayed time points must be limited with respect to the size of the polygons in order to avoid clutter.

For detecting periodic temporal patterns, Hewagamage suggested using spirals to depict time in a 3D representation similar to MacEachren's space-time cube. They used this technique to display events, i.e., data with existential changes like the sight of a bird at a specific time and place. Depending on the semantics of the data, events often show some periodic appearance patterns, e.g., bird migration depends on the season and observed birds may rest at a certain place every year. Hewagamage took a linear time line and coiled it such that one loop of the resulting three-dimensional spiral corresponded to a user-specified time interval, e.g., a year or month. At each location of interest such a spiral was positioned, and the events at that position were placed as small icons along the spiral. Thus, points in time whose temporal distance was a multiple of the selected time period were vertically aligned on the spirals. Since parts of the spirals were occluded the display needed to have interactive controls for zooming and panning, as well as for changing the period of the spirals.

References

- Geospatial-data-and-9781491984314: safaribooksonline.com, Retrieved 15 June 2018

- Surface-analysis-4348: corrosionpedia.com, Retrieved 21 July 2018

- Surface-creation-and-analysis, geoprocessing: resources.esri.com, Retrieved 05 July 2018

- GIS-and-network-analysis-23730944: researchgate.net, Retrieved 25 April 2018

- Types-of-network-analyses, network-analyst, extensions: desktop.arcgis.com, Retrieved 14 May 2018

Geospatial Topology

The science of geospatial topology is concerned with an understanding of the points, lines and polygons, which are representations of features of a geographic area. This chapter has been carefully written to provide an easy understanding of the varied facets of geospatial topology. It includes crucial topics such as spatial relation, DE-9IM, topology analysis, geography markup language, etc.

Topology

Basically, topology is the modern version of geometry, the study of all different sorts of spaces. The thing that distinguishes different kinds of geometry from each other (including topology here as a kind of geometry) is in the kinds of transformations that are allowed before you really consider something changed (This point of view was first suggested by Felix Klein, a famous German mathematician of the late 1800 and early 1900's.) In ordinary Euclidean geometry, you can move things around and flip them over, but you can't stretch or bend them. This is called "congruence" in geometry class. Two things are congruent if you can lay one on top of the other in such a way that they exactly match. In projective geometry, invented during the Renaissance to understand perspective drawing, two things are considered the same if they are both views of the same object. For example, look at a plate on a table from directly above the table, and the plate looks round, like a circle. But walk away a few feet and look at it, and it looks much wider than long, like an ellipse, because of the angle you're at. The ellipse and circle are projectively equivalent. This is one reason it is hard to learn to draw. The eye and the mind work projectively. They look at this elliptical plate on the table, and think it's a circle, because they know what happens when you look at things at an angle like that. To learn to draw, you have to learn to draw an ellipse even though your mind is saying `circle', so you can draw what you really see, instead of `what you know it is'.

In topology, any continuous change which can be continuously undone is allowed. So a circle is the same as a triangle or a square, because you just `pull on' parts of the circle to make corners and then straighten the sides, to change a circle into a square. Then you just `smooth it out' to turn it back into a circle. These two processes are continuous in the sense that during each of them, nearby points at the start are still nearby at the end, The circle isn't the same as the image of number 8, because although you can squash the middle of a circle together to make it into a the image of number 8 continuously, when you try to undo it, you have to break the connection in the middle and this is discontinuous: points that are all near the centre of the eight end up split into two batches, on opposite sides of the circle, far apart. Another example: a plate and a bowl are the same topologically, because you can just flatten the bowl into the plate. At least, this is true if you use clay which is still soft and hasn't been fired yet. Once they're fired they become Euclidean rather than topological, because you can't flatten the bowl any longer without breaking it. Topology is almost the most basic form of geometry there is. It is used in nearly all branches of mathematics in one form or another.

Geospatial Topology

In GIS, topology is implemented through data structure. An arc Info coverage is a familiar topological data structure. A coverage explicitly stores topological relationships among neighbouring polygons in the Arc Attribute Table (AAT) by storing the adjacent polygon IDs in the L Poly and R Poly fields. Adjacent lines are connected through nodes, and this information is stored in the arc-node table. The Arc Info commands, CLEAN and BUILD, enforce planar topology on data and update topology tables.

Over the past two or three decades, the general consensus in the GIS community had been that topological data structures are advantageous because they provide an automated way to handle digitizing and editing errors and artifacts, reduce data storage for polygons because boundaries between adjacent polygons are stored only once; and enable advanced spatial analyses such as adjacency, connectivity, and containment. Another important consequence of planar enforcement is that a map that has topology contains space-filling, non over lapping polygons. Consequently, so-called cartographic (i.e., non-topological) data structures are no longer used by mainstream GIS software.

Components of Topology

Topology has three basic components:

1. Connectivity (Arc – Node Topology):

- Points along an arc that define its shape are called Vertices.

- Endpoints of the arc are called Nodes.

- Arcs join only at the Nodes.

2. Area Definition / Containment (Polygon – Arc Topology):

- An enclosed polygon has a measurable area.

- Lists of arcs define boundaries and closed areas are maintained.

- Polygons are represented as a series of (x, y) coordinates that connect to define an area.

3. Contiguity:

- Every arc has a direction.

- A GIS maintains a list of Polygons on the left and right side of each arc.

- The computer then uses this information to determine which features are next to one another.

Topology in different GIS Format

Coverage

Coverage is a topology based vector data format. Coverage can be a point coverage, line coverage, or polygon coverage.

The coverage model supports three basic topological relationships:

- Connectivity: Arc connects to each other at nodes.

- Area definition: An Area is defined by a series of connected arcs.

- Contiguity: Arcs have directions and left and right polygon.

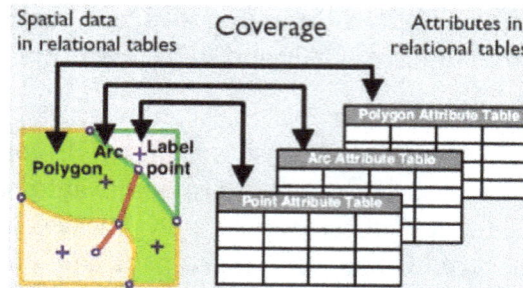

Figure: Diagram showing the coverage data structure for storing vector data.

Shapefile

Shapefile is a standard non topological data format. Shape file are a first attempt an object spatial features.They are very simple floating point geometry feature. A Shapefile is a digital vector storage format for storing geometric location and associated attribute information.

A Shapefile is actually a set of several files:

- .shp: shape format; the feature geometry itself;

- .shx: shape index format; a positional index of the feature geometry to allow seeking forwards and backwards quickly;

- .dbf: attribute format; columnar attributes for each shape, in dBase III format;

- .prj: projection format; the coordinate system and projection information, a plain text file describing the projection using well-known text format;

- .sbn: This is a binary spatial index file, which is used only by ESRI software;

- .sbx: a spatial index of the features;

- .shp.xml: metadata in XML format.

The geometry of a shapefile is stored in two basic files .shp and .shx

DXF (Drawing Exchange Format)

It maintains data in separate layers. But it does not support topology. It is AutoCAD format.

Geodatabase: A geodatabase is a relational database that store geographic information. It is Object-oriented model not a Georelational.

A relational database is a collection of tables logically associated with each other by common key attribute field.

A geodatabase can store geographic information because, besides storing a number or a string in a attribute field; tables in a geodatabase can also store geometric coordinates to define the shape and locations of points, lines or polygon.

ArcGIS supports five physical implementations of the geodatabase:

- A file geodatabase, an Access based personal geodatabase, as well as a personal, a work-group and an enterprise geodatabase.

- A personal geodatabase is file with extension .mdb, which is the file extension used by Microsoft access. A file geodatabase is folder in which file stored with .gdb extension.

	Georelational	Object based
Topological	Coverage	Geodatabase
Non-Topological	Shapefile	Geodatabase

Topological Error

Nowadays, production, storage and usage of maps in digital environments either in photogrammetric or classical methods are quite common. The most common method of producing vector maps is the precise scanning of analog maps into raster formats and then digitizing into vector forms.

During digitization process Topological Errors can be inserted in vector data.

Topological errors with polygon features can include unclosed polygons, gaps between polygon borders or overlapping polygon borders. A common topological error with polyline features is that they do not meet perfectly at a point (node). This type of error is called an under–shoot if a gap exists between the lines and an overshoot if a line ends beyond the line it should connect to.

Polygon Rules

Must be Larger than Cluster Tolerance:

Rule Description

Requires that a feature does not collapse during a validate process. This rule is mandatory for a topology and applies to all line and polygon feature classes. In instances where this rule is violated, the original geometry is left unchanged.

Potential Fixes

Delete: The Delete fix removes polygon features that would collapse during the validate process based on the topology's cluster tolerance. This fix can be applied to one or more Must Be Larger than Cluster Tolerance errors.

Examples

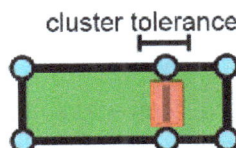

Any polygon feature, such as the one in red that would collapse when validating the topology is an error.

Must not overlap:

Rule Description

Requires the interior of polygons not overlap. The polygons can share edges or vertices. This rule is used when an area cannot belong to two or more polygons. It is useful for modelling administrative boundaries, such as ZIP Codes or voting districts, and mutually exclusive area classifications, such as land cover or landform type.

Potential Fixes

Subtract: The Subtract fix removes the overlapping portion of geometry from each feature that is causing the error and leaves a gap or void in its place. This fix can be applied to one or more selected Must Not Overlap errors.

Merge: The Merge fix adds the portion of overlap from one feature and subtracts it from the others that are violating the rule. You need to pick the feature that receives the portion of overlap using the Merge dialog box. This fix can be applied to one Must Not Overlap error only.

Create Feature: The Create Feature fix creates a new polygon feature out of the error shape and removes the portion of overlap from each of the features, causing the error to create a planar representation of the feature geometry. This fix can be applied to one or more selected Must Not Overlap errors.

Examples

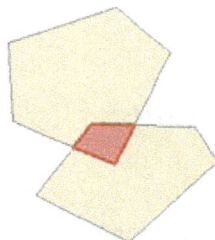

Must not have Gaps:

Rule Description

This rule requires that there are no voids within a single polygon or between adjacent polygons. All polygons must form a continuous surface. An error will always exist on the perimeter of the surface. You can either ignore this error or mark it as an exception. Use this rule on data that must completely cover an area. For example, soil polygons cannot include gaps or form voids—they must cover an entire area.

Potential Fixes

Create Feature: The Create Feature fix creates new polygon features using a closed ring of the line error shapes that form a gap. This fix can be applied to one or more selected Must Not Have Gaps errors. If you select two errors and use the Create Feature fix, the result will be one polygon feature per ring. Note that the ring that forms the outer bounds of your feature class will be in error. Using the Create Feature fix for this specific error can create overlapping polygons. Remember that you can mark this error as an exception.

Examples

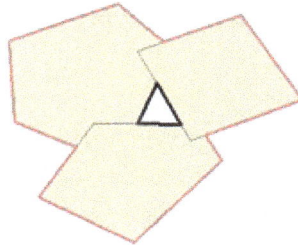

You can use Create Feature to create a new polygon in the void in the centre. You can also use Create Feature or mark the error on the outside boundary as an exception.

Must not overlap with:

Rule Description

Requires the interior of polygons in one feature class must not overlap with the interior of polygons in another feature class (or subtype). Polygons of the two feature classes can share edges or vertices or be completely disjointed. This rule is used when an area cannot belong to two separate feature classes. It is useful for combining two mutually exclusive systems of area classification, such as zoning and water body type, where areas defined within the zoning class cannot also be defined in the water body class and vice versa.

Potential Fixes

Subtract: The Subtract fix removes the overlapping portion of each feature that is causing the error and leaves a gap or void in its place. This fix can be applied to one or more selected Must Not Overlap with errors.

Merge: The Merge fix adds the portion of overlap from one feature and subtracts it from the others that are violating the rule. You need to pick the feature that receives the portion of overlap using the Merge dialog box. This fix can be applied to one Must Not Overlap With error only.

Examples

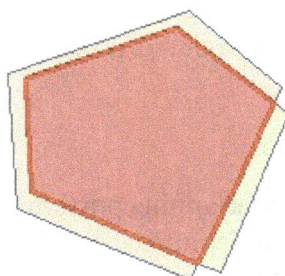

Must be Covered by Feature Class of:

Rules of Description

Requires a polygon in one feature class (or subtype) must share all of its area with polygons in another feature class (or subtype). An area in the first feature class that is not covered by polygons from the other feature class is an error. This rule is used when an area of one type, such as a state, should be completely covered by areas of another type, such as counties.

Potential Fixes

Subtract: The Subtract fix removes the overlapping portion of each feature that is causing the error so the boundary of each feature from both feature classes is the same. This fix can be applied to one or more selected Must Be Covered By Feature Class Of errors.

Create Feature: The Create Feature fix creates a new polygon feature out of the portion of overlap from the existing polygon so the boundary of each feature from both feature classes is the same. This fix can be applied to one or more selected Must Be Covered by Feature Class of errors.

Example

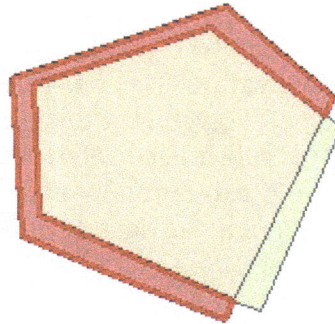

Must Cover each other:

Rules of Description

Requires the polygons of one feature class (or subtype) must share all of their area with the polygons of another feature class (or subtype). Polygons may share edges or vertices. Any area defined in either feature class that is not shared with the other is an error. This rule is used when two systems of classification are used for the same geographic area, and any given point defined in one system must also be defined in the other. One such case occurs with nested hierarchical datasets, such as census blocks and block groups or small watersheds and large drainage basins. The rule can also be applied to non-hierarchically related polygon feature classes, such as soil type and slope class.

Potential Fixes

Subtract: The Subtract fix removes the overlapping portion of each feature that is causing the error so the boundary of each feature from both feature classes is the same. This fix can be applied to one or more selected Must Cover Each Other errors.

Create Feature: The Create Feature fix creates a new polygon feature out of the portion of overlap from the existing polygon so the boundary of each feature from both feature classes is the same. This fix can be applied to one or more selected Must Cover Each Other errors.

Example

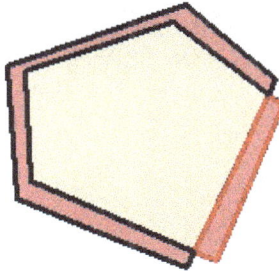

Must be Covered by:

Rules of Description

Requires polygons of one feature class (or subtype) must be contained within polygons of another feature class (or subtype). Polygons may share edges or vertices. Any area defined in the contained feature class must be covered by an area in the covering feature class. This rule is used when area features of a given type must be located within features of another type. This rule is useful when modelling areas that are subsets of a larger surrounding area, such as management units within forests or blocks within block groups.

Potential Fixes

Create Feature: The Create Feature fix creates a new polygon feature out of the portion of overlap from the existing polygon so the boundary of each feature from both feature classes is the same. This fix can be applied to one or more selected Must Be Covered By errors.

Example

Boundary must be Covered by:

Rules Description

Requires boundaries of polygon features must be covered by lines in another feature class. This rule is used when area features need to have line features that mark the boundaries of the areas. This is usually when the areas have one set of attributes and their boundaries have other attributes. For example, parcels might be stored in the geo database along with their boundaries. Each parcel might be defined by one or more line features that store information about their length or the date surveyed, and every parcel should exactly match its boundaries.

Potential Fixes

Create Feature: The Create Feature fix creates a new line feature from the boundary segments of the polygon feature generating the error. This fix can be applied to one or more selected Boundary Must Be Covered By errors.

Example

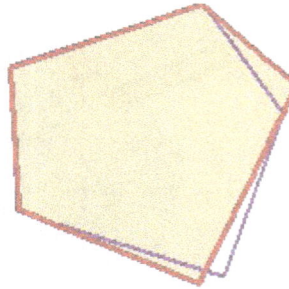

Area Boundary Must Be Covered By Boundary:

Rule Description

Requires boundaries of polygon features in one feature class (or subtype) be covered by boundaries of polygon features in another feature class (or subtype). This is useful when polygon features in one feature class, such as subdivisions, are composed of multiple polygons in another class, such as parcels, and the shared boundaries must be aligned.

Potential Fixes

None

Examples

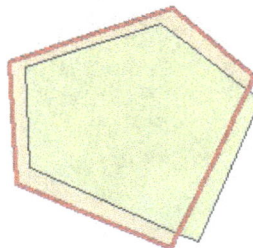

Contains Point:

Rule Description

Requires a polygon in one feature class contain at least one point from another feature class. Points must be within the polygon, not on the boundary. This is useful when every polygon should have at least one associated point, such as when parcels must have an address point.

Potential Fixes

Create Feature: The Create Feature fix creates a new point feature at the centroid of the polygon feature that is causing the error. The point feature that is created is guaranteed to be within the polygon feature. This fix can be applied to one or more selected Contains Point errors.

Examples

The top polygon is an error because it does not contain a point.

Contains One Point:

Rule Description

Requires each polygon contains one point feature and that each point feature falls within a single polygon. This is used when there must be a one-to-one correspondence between features of a polygon feature class and features of a point feature class, such as administrative boundaries and their capital cities. Each point must be properly inside exactly one polygon and each polygon must properly contain exactly one point. Points must be within the polygon, not on the boundary.

Potential Fixes

None

Examples

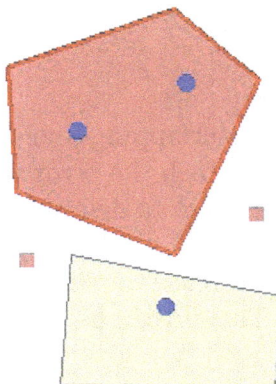

The top polygon is an error because it contains more than one point. Points are errors when they are outside a polygon.

Line Rules

Must be Larger than Cluster Tolerance:

Rule Description

Requires that a feature does not collapse during a validate process. This rule is mandatory for a topology and applies to all line and polygon feature classes. In instances where this rule is violated, the original geometry is left unchanged.

Potential Fixes

Delete: The Delete fix removes line features that would collapse during the validate process based on the topology's cluster tolerance. This fix can be applied to one or more Must Be Larger than Cluster Tolerance errors.

Examples

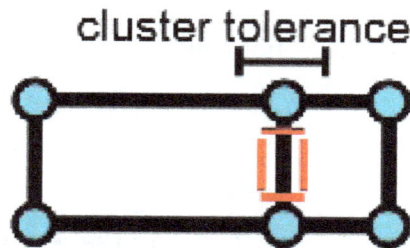

Any line feature, such as these lines in red that would collapse when validating the topology is an error.

Must Not Overlap:

Rules of Description

Requires lines not overlap with lines in the same feature class (or subtype). This rule is used where line segments should not be duplicated, for example, in a stream feature class. Lines can cross or intersect but cannot share segments.

Potential Fixes

Subtract: The Subtract fix removes the overlapping line segments from the feature causing the error. You must select the feature from which the error will be removed. If you have duplicate line features, select the line feature you want to delete from the Subtract dialog box. Note that the Subtract fix will create multipart features, so if the overlapping segments are not at the end or start of a line feature, you might want to use the Explode command on the Advanced Editing toolbar to create single-part features. This fix can be applied to one selected Must Not Overlap error only.

Examples

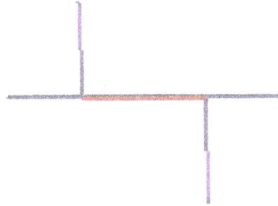

Must not Intersect:

Rule Description

Requires that line features from the same feature class (or subtype) not cross or overlap each other. Lines can share endpoints. This rule is used for contour lines that should never cross each other or in cases where the intersection of lines should only occur at endpoints, such as street segments and intersections.

Potential Fixes

Subtract: The Subtract fix removes the overlapping line segments from the feature causing the error. You must select the feature from which the error will be removed. If you have duplicate line features, select the line feature you want to delete from the Subtract dialog box. Note that the Subtract fix will create multipart features, so if the overlapping segments are not at the end or start of a line feature, you might want to use the Explode command on the Advanced Editing toolbar to create single-part features. This fix can be applied to one Must Not Intersect error only.

Split: The Split fix splits the line features that cross one another at their point of intersection. If two lines cross at a single point, applying the Split fix at that location will result in four features. Attributes from the original features will be maintained in the split features. If a split policy is present, the attributes will be updated accordingly. This fix can be applied to one or more Must Not Intersect errors.

Example

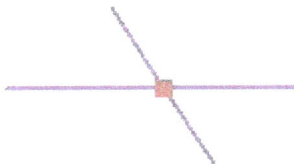

Must not Intersect with:

Rule Description

Requires that line features from one feature class (or subtype) not cross or overlap lines from another feature class (or subtype). Lines can share endpoints. This rule is used when there are lines from two layers that should never cross each other or in cases where the intersection of lines should only occur at endpoints, such as streets and railroads.

Potential Fixes

Subtract: The Subtract fix removes the overlapping line segments from the feature causing the error. You must select the feature from which the error will be removed. If you have duplicate line features, select the line feature you want to delete from the Subtract dialog box. Note that the Subtract fix will create multipart features, so if the overlapping segments are not at the end or start of a line feature, you might want to use the Explode command on the Advanced Editing toolbar to create single-part features. This fix can be applied to one Must Not Intersect with error only.

Split: The Split fix splits the line features that cross one another at their point of intersection. If two lines cross at a single point, applying the Split fix at that location will result in four features. Attributes from the original features will be maintained in the split features. If a split policy is present, the attributes will be updated accordingly. This fix can be applied to one or more Must Not Intersect with errors.

Example

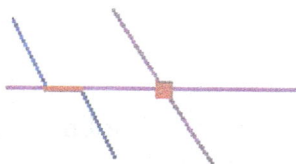

Must not have Dangles:

Rule Description

Requires a line feature must touch lines from the same feature class (or subtype) at both end-points. An endpoint that is not connected to another line is called a dangle. This rule is used when line features must form closed loops, such as when they are defining the boundaries of polygon features. It may also be used in cases where lines typically connect to other lines, as with streets. In this case, exceptions can be used where the rule is occasionally violated, as with cul-de-sac or dead-end street segments.

Potential Fixes

Extend: The Extend fix will extend the dangling end of line features if they snap to other line features within a given distance. If no feature is found within the distance specified, the feature will not extend to the distance specified. Also, if several errors were selected, the fix will simply skip the features that it cannot extend and attempt to extend the next feature in the list. The errors of features that could not be extended remain on the Error Inspector dialog box. If the distance value is 0, lines will extend until they find a feature to snap to. This fix can be applied to one or more Must Not Have Dangles errors.

Trim: The Trim fix will trim dangling line features if a point of intersection is found within a given distance. If no feature is found within the distance specified, the feature will not be trimmed, nor will it be deleted if the distance is greater than the length of the feature in error. If the distance value is 0, lines will be trimmed back until they find a point of intersection. If no intersection is located, the feature will not be trimmed and the fix will attempt to trim the next feature in error. This fix can be applied to one or more Must Not Have dangles errors.

Snap: The Snap fix will snap dangling line features to the nearest line feature within a given distance. If no line feature is found within the distance specified, the line will not be snapped. The Snap fix will snap to the nearest feature found within the distance. It searches for endpoints to snap to first, then vertices, and finally to the edge of line features within the feature class. The Snap fix can be applied to one or more Must Not Have Dangles errors.

Example

Must not have Pseudo Nodes:

Rule of Description

Requires a line connect to at least two other lines at each endpoint. Lines that connect to one other line (or to themselves) are said to have pseudo nodes. This rule is used where line features must form closed loops, such as when they define the boundaries of polygons or when line features logically must connect to two other line features at each end, as with segments in a stream network, with exceptions being marked for the originating ends of first-order streams.

Potential Fixes

Merge To Largest: The Merge to largest fix will merge the geometry of the shorter line into the geometry of the longest line. The attributes of the longest line feature will be retained. This fix can be applied to one or more Must Not Have Pseudo Nodes errors.

Merge: The Merge fix adds the geometry of one line feature into the other line feature causing the error. You must pick the line feature into which to merge. This fix can be applied to one selected Must Not Have Pseudo Nodes error.

Example

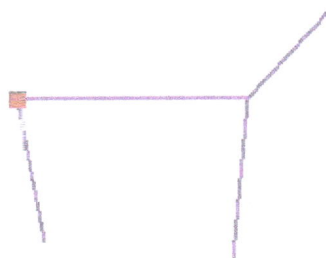

Must not Intersect or Touch Interior:

Rule of Description

Requires a line in one feature class (or subtype) must only touch other lines of the same feature class (or subtype) at endpoints. Any line segment in which features overlap or any intersection not at an endpoint is an error. This rule is useful where lines must only be connected at endpoints, such as in the case of lot lines, which must split (only connect to the endpoints of) back lot lines and cannot overlap each other.

Potential Fixes

Subtract: The Subtract fix removes the overlapping line segments from the feature causing the error. You must select the feature from which the error will be removed. If you have duplicate line features, choose the line feature you want to delete from the Subtract dialog box. The Subtract fix creates multipart features, so if the overlapping segments are not at the end or start of a line feature, you might want to use the Explode command on the Advanced Editing toolbar to create single-part features. This fix can be applied to one selected Must Not Intersect Or Touch Interior error only.

Split: The Split fix splits the line features that cross one another at their point of intersection. If two lines cross at a single point, applying the Split fix at that location will result in four features. Attributes from the original features will be maintained in the split features. If a split policy is present, the attributes will be updated accordingly. This fix can be applied to one or more Must Not Intersect or touch Interior errors.

Example

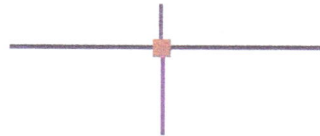

Must not Intersect or Touch Interior with:

Rule Description

Requires a line in one feature class (or subtype) must only touch other lines of another feature class (or subtype) at endpoints. Any line segment in which features overlap or any intersection not at an endpoint is an error. This rule is useful where lines from two layers must only be connected at endpoints.

Potential Fixes

Subtract: The Subtract fix removes the overlapping line segments from the feature causing the error. You must select the feature from which the error will be removed. If you have duplicate line features, choose the line feature you want to delete from the Subtract dialog box. The Subtract fix creates multipart features, so if the overlapping segments are not at the end or start of a line feature, you might want to use the Explode command on the Advanced Editing toolbar to create single-part features. This fix can be applied to one selected Must Not Intersect Or Touch Interior With error only.

Split: The Split fix splits the line features that cross one another at their point of intersection. If two lines cross at a single point, applying the Split fix at that location will result in four features. Attributes from the original features will be maintained in the split features. If a split policy is present, the attributes will be updated accordingly. This fix can be applied to one or more Must Not Intersect Or Touch Interior With errors.

Example

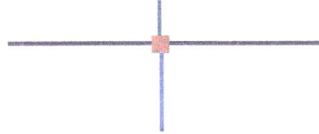

Must not Overlap with:

Rule Description

Requires a line from one feature class (or subtype) not overlap with line features in another feature class (or subtype). This rule is used when line features cannot share the same space. For example, roads must not overlap with railroads or depression subtypes of contour lines cannot overlap with other contour lines.

Potential Fixes

Subtract: The Subtract fix removes the overlapping line segments from the feature causing the error. You must select the feature from which the error will be removed. If you have duplicate line features, choose the line feature you want to delete from the Subtract dialog box. The Subtract fix creates multipart features, so if the overlapping segments are not at the end or start of a line feature, you might want to use the Explode command on the Advanced Editing toolbar to create single-part features. This fix can be applied to one selected Must Not Overlap With error only.

Example

Where the purple lines overlap is an error.

Must be Covered by Feature Class of:

Rule of Description

Requires a lines from one feature class (or subtype) must be covered by the lines in another feature class (or subtype). This is useful for modelling logically different but spatially coincident lines, such as routes and streets. A bus route feature class must not depart from the streets defined in the street feature class.

Potential Fixes

None

Example

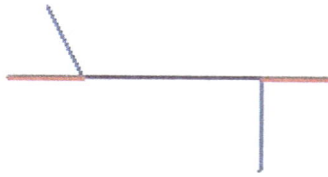

Where, the purple lines do not overlap is an error.

Must be Covered by Boundary of:

Rule of Description

Requires lines be covered by the boundaries of area features. This is useful for modelling lines, such as lot lines, that must coincide with the edge of polygon features, such as lots.

Potential Fixes

Subtract: The Subtract fix removes line segments that are not coincident with the boundary of polygon features. If the line feature does not share any segments in common with the boundary of a polygon feature, the feature will be deleted. This fix can be applied to one or more Must Be Covered By Boundary Of errors.

Example

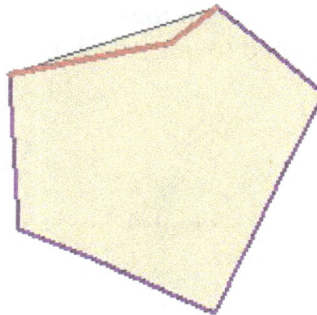

Must be Inside:

Rule of Description

Requires a line is contained within the boundary of an area feature. This is useful for cases where lines may partially or totally coincide with area boundaries but cannot extend beyond polygons, such as state highways that must be inside state borders and rivers that must be within water sheds.

Potential Fixes

Delete: The Delete fix removes line features that are not within polygon features. Note that you can use the Edit tool and move the line inside the polygon feature if you do not want to delete it. This fix can be applied to one or more Must Be Inside errors.

Example

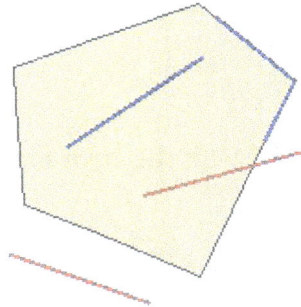

Endpoint must be Covered by:

Rule of Description

Requires the endpoints of line features must be covered by point features in another feature class. This is useful for modelling cases where a fitting must connect two pipes or a street intersection must be found at the junction of two streets.

Potential Fixes

Create Feature: The Create Feature fix adds a new point feature at the endpoint of the line feature that is in error. The Create Feature fix can be applied to one or more Endpoint Must Be Covered By errors.

Example

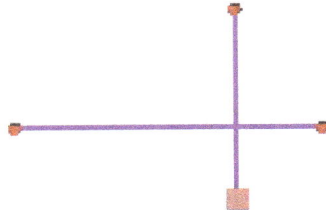

The square at the bottom indicates an error, because there is no point covering the endpoint of the line.

Must not Self-overlap:

Rule of Description

Requires that line features not overlap themselves. They can cross or touch themselves but must not have coincident segments. This rule is useful for features, such as streets, where segments might touch in a loop but where the same street should not follow the same course twice.

Potential Fixes

Simplify: The Simplify fix removes self-overlapping line segments from the feature in error. Applying the Simplify fix can result in multipart features, which you can detect using the Must Be Single Part rule. The Simplify fix can be applied to one or more Must Not Self-Overlap errors.

Example

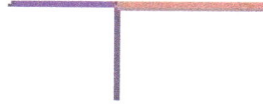

The individual line feature overlaps itself, with the error indicated by the coral line.

Must not Self-intersect

Rule Description

Requires that line features not cross or overlap themselves. This rule is useful for lines, such as contour lines, that cannot cross themselves.

Potential Fixes

Simplify: The Simplify fix removes self-overlapping line segments from the feature in error. Note that applying the Simplify fix can result in multipart features. You can detect multipart features using the Must Be Single Part rule. This fix can be applied to one or more Must Not Self-Intersect errors.

Example

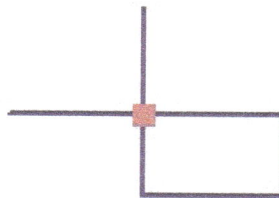

Must be Single Part

Rule Description

Requires lines have only one part. This rule is useful where, line features, such as highways may not have multiple parts.

Potential Fixes

Explode: The Explode fix creates single-part line features from each part of the multipart line feature that is in error. This fix can be applied to one or more Must Be Single Part errors.

Example

Multipart lines are created from a single sketch.

Point Rules

Must Coincide with:

Rule Description

Requires points in one feature class (or subtype) be coincident with points in another feature class (or subtype). This is useful for cases where points must be covered by other points, such as transformers must coincide with power poles in electric distribution networks and observation points must coincide with stations.

Potential Fixes

Snap: The Snap fix will move a point feature in the first feature class or subtype to the nearest point in the second feature class or subtype that is located within a given distance. If no point feature is found within the tolerance specified, the point will not be snapped. The Snap fix can be applied to one or more Must Coincide With errors.

Examples

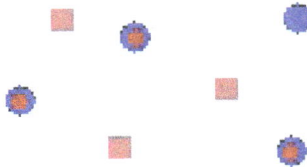

Where, a red point is not coincident with a blue point is an error.

Must be Disjoint

Rule Description

Requires points be separated spatially from other points in the same feature class (or subtype). Any points that overlap are errors. This is useful for ensuring that points are not coincident or duplicated within the same feature class, such as in layers of cities, parcel lot ID points, wells, or streetlamp poles.

Potential Fixes

None

Example

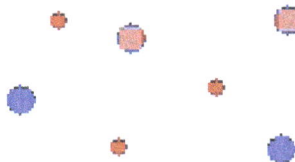

Where, a red point and a blue point overlap is an error.

Must be Covered by Boundary of:

Rule Description

Requires points fall on the boundaries of area features. This is useful when the point features help support the boundary system, such as boundary markers, which must be found on the edges of certain areas.

Potential Fixes

None

Example

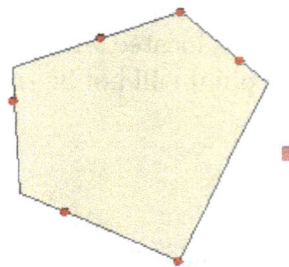

The square on the right indicates an error because it is a point that is not on the boundary of the polygon.

Must be Properly Inside

Rule Description

Requires points fall within area features. This is useful when the point features are related to polygons, such as wells and well pads or address points and parcels.

Potential Fixes

Delete: The Delete fix removes point features that are not properly within polygon features. Note that you can use the Edit tool and move the point inside the polygon feature if you do not want to delete it. This fix can be applied to one or more Must Be Properly Inside errors.

Example

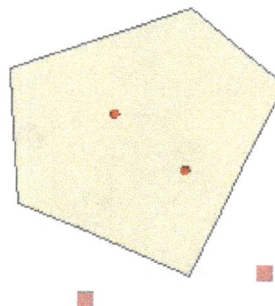

The squares are errors where there are points that are not inside the polygon.

Must be Covered by Endpoint of:

Rule of Description

Requires points in one feature class must be covered by the endpoints of lines in another feature class. This rule is similar to the line rule Endpoint Must Be Covered By except that, in cases where the rule is violated, it is the point feature that is marked as an error rather than the line. Boundary corner markers might be constrained to be covered by the endpoints of boundary lines.

Potential Fixes

Delete: The Delete fix removes point features that are not coincident with the endpoint of line features. Note that you can snap the point to the line by setting edge snapping to the line layer, then moving the point with the Edit tool. This fix can be applied to one or more Must Be Covered by Endpoint of errors.

Example

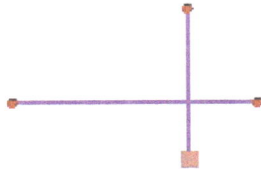

The square indicates an error where the point is not on an endpoint of a line.

Must be Covered by Line:

Rule Description

Requires points in one feature class be covered by lines in another feature class. It does not constrain the covering portion of the line to be an endpoint. This rule is useful for points that fall along a set of lines, such as highway signs along highways.

Potential Fixes

Example

The squares are points that are not covered by the line.

Topological Digitizing and Editing

One of the primary reasons topology was developed was to provide a rigorous, automated method to clean up data entry errors and verify data. The typical digitizing procedure is to digitize all lines, build topology, and label polygons and then clean up slivers, dangles, and under- and overshoots and build topology again, repeating the clean and build phases as many times as necessary.

Editing, what if the process did not start with the tangled mess of "cartographic spaghetti"? By approaching digitizing from a feature-centric perspective and enforcing planar topology when each feature boundary is digitized and labelled, sliver polygons, dangling nodes, missing labels and multi-labelled features would be eliminated. To be fair, computer hardware was not always powerful enough to support a feature-centric digitizing approach that requires on-the-fly calculation of geometric intersections.

Today's computers are powerful enough to support feature-centric digitizing for most GIS users. ArcView GIS supports such feature-centric digitizing through the Append Polygon, Split Polygon, and Split Line tools. With these tools users can add a polygon (or line) adjacent to an existing polygon and have boundaries match perfectly. ArcView GIS also supports topological editing of shared boundaries or nodes through the manipulation of vertices.

File Sizes no Longer an Issue

A second oft-cited advantage of topological data structures is smaller file sizes because shared vertices of adjacent polygons are not stored twice. Theoretically these files should be up to half the size of non-topological files. In practice however, shape files are rarely twice as large as the same data stored in coverages, in part because coverages require additional files to store the topological information. Attribute tables are often a large proportion of the overall file size but are the same size regardless of how feature geometry is stored. Moreover, although storage was often an important consideration in the past, the current low cost of storage means that for most GIS user storage space is not a constraint.

Finding Adjacent Features

Perhaps the most pervasive misunderstanding about shape files is that because topology is not explicitly stored, adjacent features cannot be found. However, adjacent features can easily be found by intersecting target polygons with other polygons in the same map and identifying the points of intersection of polygons that touch boundaries or overlap. The geometric intersections of adjacent features are calculated on the fly by comparing the vertices of adjacent features rather than looking up adjacent features in a table.

For example, to find all the neighbouring parcels of a parcel, select the parcel, choose Theme > Select by Theme from the View menu and choose "intersect" from the drop-down box and click on New Set to select all the parcels immediately adjacent to the originally selected parcel. More complex adjacency analyses can be accomplished by combining the selection by theme with a query for specific attributes such as identifying only the residential parcels adjacent to an industrial parcel. While some of the more complex adjacencies that involve direction (e.g., find the adjacent parcels to the east of a given road) are much more difficult to accomplish without stored topology, these analyses are not frequent nor are they a make-it-or-break-it requirement for typical users.

Computing Adjacency Lists

Although analytical operations that require adjacency information can be performed in ArcView GIS through the interface, performance requirements many necessitate building a table to store adjacency information. Two algorithms for building lists of adjacent features, described here, could be incorporated in an Avenue script. Although the representation of topological spatial relationships traditionally has been restricted to exactly adjacent neighbours, this restriction can be relaxed to find adjacent features. The notion of adjacency can be extended to include features that are within some distance (D) rather than exactly adjacent (D = 0). One advantage of computing adjacency lists is that adjacency can be defined in relation to the spatial precision of the coordinates, making analysis less sensitive to sliver polygons. These algorithms can find adjacency for polylines and polygons. If D > 0, adjacent points can also be identified.

One algorithm creates an adjacency list using the so-called "brute-force" approach. In this simple algorithm, for every pair of features, it determines if these features intersect and stores the adjacent index values. The time required for this algorithm is proportional to the square of the number of features (N) or order $O(N2)$. However, because adjacency is a reflexive spatial relation, the brute-force algorithm can be modified to store the index for reciprocal features as well. The time required for the modified algorithm is proportional to $O(N(N-1)/2)$. A second algorithm uses a "divide and conquer" approach to recursively subdivide features into smaller and smaller groups. The reflexive brute-force approach is applied to these smaller groups. Because the number of features in each group is smaller than N, the overall number of intersection tests between two features is reduced considerably.

Checking Topology in Shape files

A planar-enforced shape file can be created as described above or derived from coverage. However, if non-topological editing methods are used, a shape file can lose its planar topology during editing. Planar topology can be enforced on shape files with the assistance of some Avenue scripts.

The first step in enforcing planar topology in a shape-file is to remove twisted or self-intersecting polygon rings and to ensure that the "inside" of the polygon is on the correct side of the polygon boundary. Next, gaps are identified by creating a rectangle that encompasses all the polygons of interest and serves as a backdrop. The polygons are subtracted from the rectangle containing all polygons. The remaining areas are gaps. A gap polygon is removed by merging it into an adjacent polygon or by making it a legitimate polygon. Overlaps are found by intersecting each polygon with all other polygons. If an intersection is found, then the polygon representing the overlap is created. Overlaps can be removed by deleting the overlapping area from one of the involved polygons. Once boundary changes have been made, the area and perimeter of each polygon should be recalculated.

In conclusion, the standard notion of topology in GIS centres around explicit representation of adjacent spatial relations and involves planar enforcement of geographic features. Although shape files do not explicitly store spatial relations, they can conform to planar enforcement. If, during map production or editing, planar enforcement is violated, then statistical summations that assume space-filling polygons could be inaccurate.

Although this may be heresy to many users, there are advantages to using shape files that violate planar assumptions (i.e., shape files that have overlaps and/or gaps). Many useful analyses do not require data with precise planar topology, but these analyses are never conducted because it is assumed that base data must have topology. For instance, city and county governments find it extremely time-consuming and difficult to build parcel coverages because parcel boundary descriptions rarely match cleanly with adjacent parcels. Resolving boundary disputes is a very time-consuming process, often fraught with complicated legal issues. However, a standard query of parcel data performed with reasonably coincident boundaries (i.e., sub meter accuracy) can be used to find landowners within a certain distance of a given location for notification purposes.

Though the advantages previously attributed to topological data structures have become less clear, in large part because of improvements in computer performance, the bottom line is that GIS users need to adequately understand the data structures and use them appropriately.

Spatial Relation

A primary function of a geographic information system is determining the spatial relationships between features. The distance separating a hazardous waste disposal site and hospital, school or housing development is an example of a spatial relationship.

Predicates are Boolean functions that return TRUE if a test passes and FALSE, otherwise, to determine if a specific relationship exists between a pair of geometries. Other functions return a value as a result of a spatial relationship. The result returned by the distance function, the space separating two geometries, is a double precision number. Alternatively, functions like intersection return geometry as the result of combining two geometries.

Predicates

Predicates return t (TRUE) if a comparison meets the functions criteria; otherwise, they return f (FALSE). Predicates that test for a spatial relationship compare pairs of geometry that can be a different type or dimension.

Predicates compare the X and Y coordinates of the submitted geometries. The Z coordinates and measure values, if they exist, are ignored. Geometries that have Z coordinates or measures can be compared with those that don't.

The Dimensionally Extended 9 Intersection Model (DE-9IM) developed by Clementini, et al., dimensionally extends the 9 Intersection Model of Egenhofer and Herring. DE-9IM is a mathematical approach that defines the pair-wise spatial relationship between geometries of different types and dimensions. This model expresses spatial relationships among all types of geometry as pair-wise intersections of their interior, boundary, and exterior with consideration for the dimension of the resulting intersections.

Given geometries a and b, I(a), B(a), and E(a) represent the interior, boundary, and exterior of a, and I(b), B(b), and E(b) represent the interior, boundary, and exterior of b. The intersections of I(a), B(a), and E(a) with I(b), B(b), and E(b) produces a 3-by-3 matrix. Each intersection can result in geometries of different dimensions. For example, the intersection of the boundaries of two polygons could consist of a point and a linestring, in which case the dim function would return the maximum dimension of 1.

The dim function returns a value of -1, 0, 1, or 2. The -1 corresponds to the null set that is returned when no intersection was found or dim(Æ).

	Interior	Boundary	Exterior
Interior	dim(I(a) ∩ I(b))	dim(I(a) ∩ B(b))	dim(I(a) ∩ E(b))
Boundary	dim(B(a) ∩ I(b))	dim(B(a) ∩ B(b))	dim(B(a) ∩ E(b))
Exterior	dim(E(a) ∩ I(b))	dim(E(a) ∩ B(b))	dim(E(a) ∩ E(b))

The results of the spatial relationship predicates can be understood or verified by comparing the results of the predicate with a pattern matrix that represents the acceptable values for the DE-9IM.

The pattern matrix contains the acceptable values for each of the intersection matrix cells. The possible pattern values are:

TAn intersection must exist; dim = 0, 1, or 2.

FAn intersection must not exist; dim = -1.

It does not matter if an intersection exists or not; dim = -1, 0, 1, or 2.

0An intersection must exist and its maximum dimension must be 0; dim = 0.

1An intersection must exist and its maximum dimension must be 1; dim = 1.

2An intersection must exist and its maximum dimension must be 2; dim = 2.

Each predicate has at least one pattern matrix, but some require more than one to describe the relationships of various geometry type combinations.

The pattern matrix of the within predicate for geometry combinations has the following form:

b		Interior	Boundary	Exterior
	Interior	T	*	F
a	Boundary	*	*	F
	Exterior	*	*	*

Simply put, the within predicate returns true when the interiors of both geometries intersect, and the interior and boundary of a does not intersect the exterior of b. All other conditions do not matter.

Equal

Equal returns t (TRUE) if two geometries of the same type have identical X,Y coordinate values.

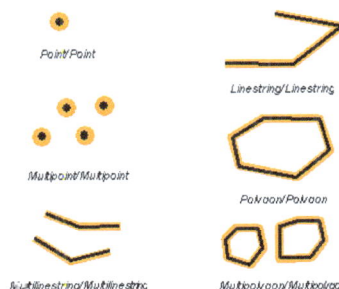

Point/Point

Linestring/Linestring

Multipoint/Multipoint

Polygon/Polygon

Multilinestring/Multilinestring

Multipolygon/Multipolygon

Geometries are equal if they have matching X,Y coordinates.

The DE-9IM pattern matrix for equality ensures that the interiors intersect and that no part interior or boundary of either geometry intersects the exterior of the other.

		b		
		Interior	Boundary	Exterior
a	Interior	T	*	F
	Boundary	*	*	F
	Exterior	F	F	*

Disjoint

Disjoint returns t (TRUE) if the intersection of the two geometries is an empty set.

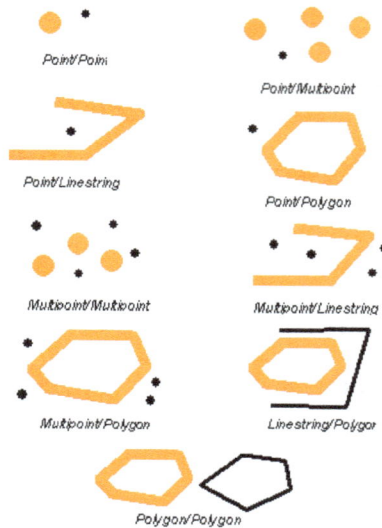

Point/Point

Point/Multipoint

Point/Linestring

Point/Polygon

Multipoint/Multipoint

Multipoint/Linestring

Multipoint/Polygon

Linestring/Polygon

Polygon/Polygon

Geometries are disjoint if they do not intersect one another in any way.

The disjoint predicates pattern matrix simply states that neither the interiors nor the boundaries of either geometry intersect.

		b		
		Interior	Boundary	Exterior
a	Interior	F	F	*
	Boundary	F	F	*
	Exterior	*	*	*

Intersects

Intersects returns t (TRUE) if the intersection does not result in an empty set. Intersects returns the exact opposite result of disjoint.

The intersect predicate will return TRUE if the conditions of any of the following pattern matrices returns TRUE.

The intersect predicate returns TRUE if the interiors of both geometries intersect.

		b		
		Interior	Boundary	Exterior
a	Interior	T	*	*
	Boundary	*	*	*
	Exterior	*	*	*

The intersect predicate returns TRUE if the boundary of the first geometry intersects the boundary of the second geometry.

		b		
		Interior	Boundary	Exterior
a	Interior	*	T	*
	Boundary	*	*	*
	Exterior	*	*	*

The intersects predicate returns TRUE if the boundary of the first geometry intersects the Interior of the second.

		b		
		Interior	Boundary	Exterior
a	Interior	*	*	*
	Boundary	T	*	*
	Exterior	*	*	*

The intersect predicate returns TRUE if the boundaries of either geometry intersect.

		b		
		Interior	Boundary	Exterior
a	Interior	*	*	*
	Boundary	*	T	*
	Exterior	*	*	*

Touch

Touch returns t (TRUE) if none of the point common to both geometries intersects the interiors of both geometries. At least one geometry must be a line string, polygon, multiline string or multi polygon be line string, polygon, multiline string or multi polygon.

Touch returns TRUE if either of the geometries boundaries intersect, if only one of the geometries interiors intersects the others boundary.

The pattern matrices show us that the touch predicate returns TRUE when the interiors of the geometry don't intersect and the boundary of either geometry intersects the others interior or boundary.

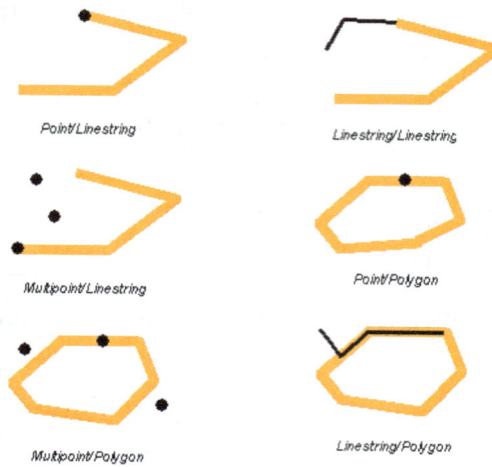

The touch predicate returns TRUE if the boundary of one geometry intersects the interior of the other but the interiors do not intersect.

		b		
		Interior	Boundary	Exterior
a	Interior	F	T	*
	Boundary	*	*	*
	Exterior	*	*	*

The touch predicate returns TRUE if the boundary of one geometry intersects the interior of the other but the interiors do not intersect.

		b		
		Interior	Boundary	Exterior
a	Interior	F	*	*
	Boundary	T	*	*
	Exterior	*	*	*

The touch predicate returns TRUE if the boundaries of both geometries intersect but the interiors do not.

		b		
		Interior	Boundary	Exterior
a	Interior	F	*	*
	Boundary	*	T	*
	Exterior	*	*	*

Overlap

Overlap compares two geometries of the same dimension and returns t (TRUE) if their intersection set results in geometry different from both but of the same dimension.

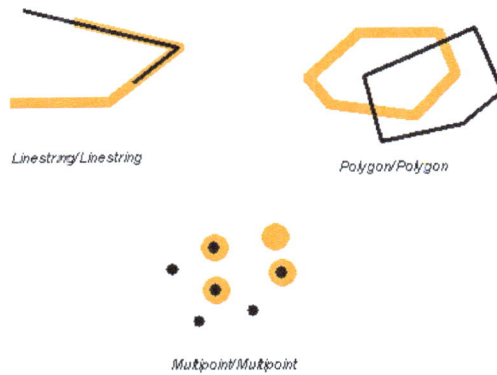

Linestring/Linestring

Polygon/Polygon

Multipoint/Multipoint

Overlap returns t (TRUE) only for geometries of the same dimension and only when their intersection set results in a geometry of the same dimension. In other words, if the intersection of two polygons results in polygon, then overlap returns t (TRUE).

This pattern matrix applies to polygon/polygon, multipoint/multipoint and multi polygon/multi polygon overlays. For these combinations the overlap predicate returns TRUE if the interior of both geometries intersects the others interior and exterior.

		b		
		Interior	Boundary	Exterior
	Interior	T	*	T
a	Boundary	*	*	*
	Exterior	T	*	*

This pattern matrix applies to line string/line string and multiline string/multiline string overlaps. In this case the intersection of the geometries must result in a geometry that has a dimension of 1 (another line string). If the dimension of the intersection of the interiors had resulted in 0 (a point) the overlap predicate would return FALSE; however, the cross predicate would have returned TRUE.

		b		
		Interior	Boundary	Exterior
	Interior	1	*	T
a	Boundary	*	*	*
	Exterior	T	*	*

Cross

Cross returns t (TRUE) if the intersection results in a geometry, whose dimension is one less than maximum dimension of the two source geometries and the intersection set is interior to both source geometries. Cross returns t (TRUE) for only multipoint/polygon, multipoint/line string, line string/line string, line string/polygon and line string/multi polygon comparisons.

Cross returns t (TRUE) if the dimension of the intersection is one less than the maximum dimension of the source geometries and the interiors of both geometries are intersected.

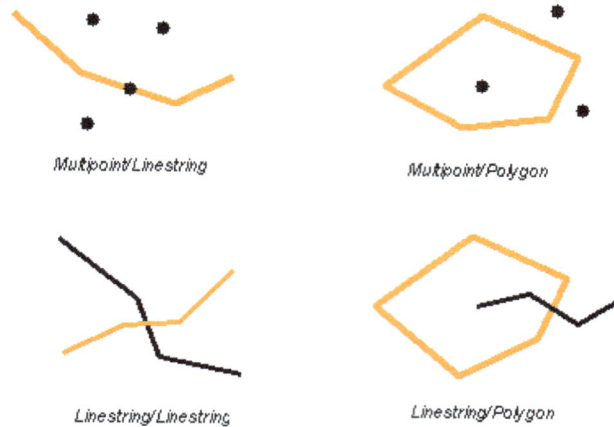

Multipoint/Linestring   Multipoint/Polygon

Linestring/Linestring   Linestring/Polygon

This cross predicate pattern matrix applies to multipoint/line string, multipoint/multiline string, multipoint/polygon, multipoint/multi-polygon, line string/polygon and line string/multi-polygon. The matrix states that the interiors must intersect and that at least the interior of the primary (geometry a) must intersect the exterior of the secondary (geometry b).

		b		
		Interior	Boundary	Exterior
	Interior	T	*	T
a	Boundary	*	*	*
	Exterior	*	*	*

This cross predicate matrix applies to line string/line string, line string/multiline string, and multiline string/multiline string. The matrix states that the dimension of the intersection of the interior must be 0 (intersect at a point). If the dimension of this intersection was 1 (intersect at a line string) the cross predicate would return FALSE but the overlap predicate would return TRUE.

		b		
		Interior	Boundary	Exterior
	Interior	0	*	*
a	Boundary	*	*	*
	Exterior	*	*	*

Within

Within returns t (TRUE) if the first geometry is completely within the second geometry. Within tests for the exact opposite result of contains.

Within returns t (TRUE) if the first geometry is completely inside the second geometry. The boundary and interior of the first geometry are not allowed to intersect the exterior of the second geometry and the first geometry may not equal the second geometry.

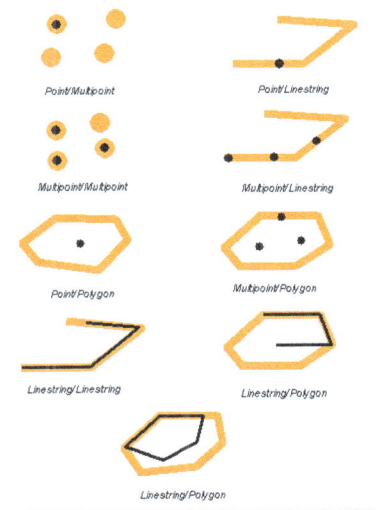

The within predicate pattern matrix states that the interiors of both geometries must intersect and that the interior and boundary of the primary geometry (geometry a) must not intersect the exterior of the secondary (geometry b).

		b		
		Interior	Boundary	Exterior
a	Interior	T	*	F
	Boundary	*	*	F
	Exterior	*	*	*

Contains

Contains returns t (TRUE) if the second geometry is completely contained by the first geometry. The contain predicate returns the exact opposite result of the within predicate.

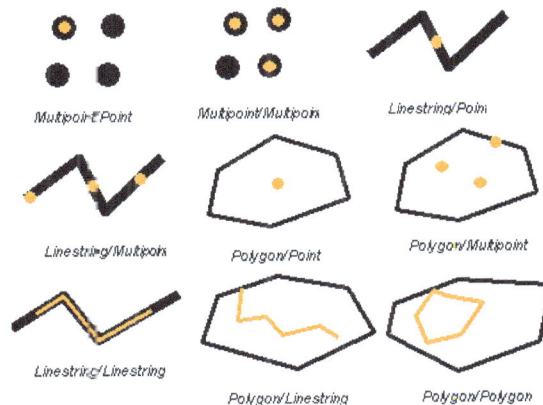

Contains returns t (TRUE) if the second geometry is completely inside the first. The boundary and interior of the second geometry are not allowed to intersect the exterior of the first geometry and the geometries may not be equal.

The pattern matrix of the contains predicate states that the interiors of both geometries must intersect and that the interior and boundary of the secondary (geometry b) must not intersect the exterior of the primary (geometry a).

		b		
		Interior	Boundary	Exterior
a	Interior	T	*	*
	Boundary	*	*	*
	Exterior	F	F	*

Minimum Distance

The minimum distance separating disjoint features could represent the shortest distance an aircraft must travel between two locations. The distance function reports the minimum distance separating two disjoint features. If the features are not disjoint the function will report a zero minimum distance.

Intersection of Geometries

The intersection function returns the intersection set of two geometries. The intersection set is always returned as a collection that is the minimum dimension of the source geometries. For example, for a line string that intersects a polygon, the intersection function returns that portion of the line string common to the interior and boundary of the polygon as a multiline string. The multiline string contains more than one line string if the source line string intersected the polygon with two or more discontinuous segments. If the geometries do not intersect or if the intersection results in a dimension less than both source geometries, an empty geometry is returned. The following figure illustrates some examples of the intersection function.

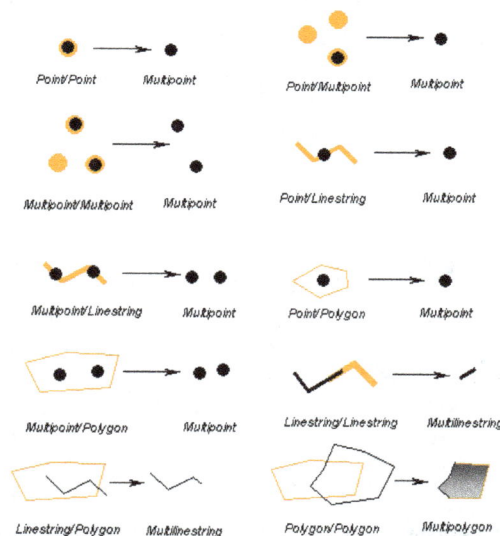

The intersection function returns the intersection set as the geometry that is the minimum dimension of the source geometries.

Difference of Geometries

The difference function returns the portion of the primary geometry that is not intersected by the secondary geometry the logical AND NOT of space. The difference function only operates on geometries of like dimension and returns a collection that has the same dimension as the source geometries. In the event that the source geometries are equal, an empty geometry is returned.

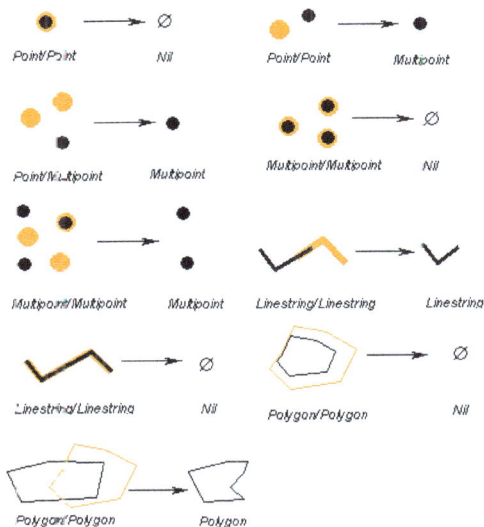

Difference returns that portion of the first geometry that is not intersected by the second.

Union of Geometries

The union function returns the union set of two geometries the Boolean logical OR of space. The source geometries must have the same dimension. Union always returns the result as a collection.

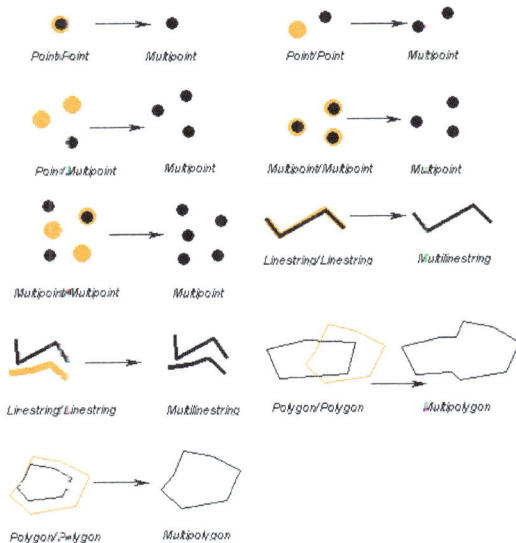

Union returns the union set of two geometries.

Symmetric Difference of Geometries

The symmetric function returns the symmetric difference of two geometries the logical XOR of space. The source geometries must have the same dimension. If the geometries are equal, the symmetric function returns an empty geometry; otherwise, the function returns the result as a collection.

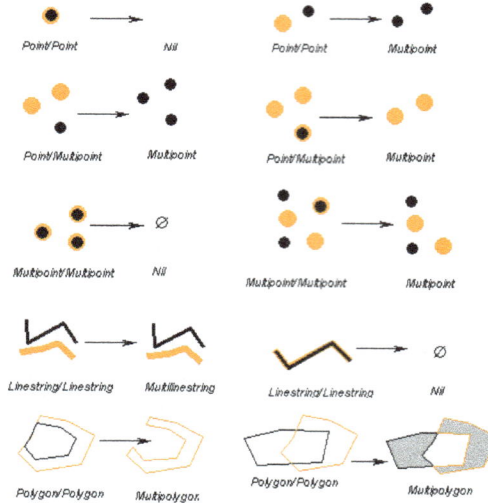

Symmetric returns the portions of the source geometries that are not part of the intersection set. The source geometries must be of the same dimension.

The combinations of geometry types and the spatial relationships that can be used are listed in the following table:

Feature class 1	Feature class 2	Compatible spatial relationship types
Point	Point	• Contains • Intersects • Relation • Within
Point	Multipoint	• Intersects • Relation • Touches
Point	Line	• Intersects • Relation • Touches • Within
Point	Area	• Intersects • Relation • Touches • Within

Feature class 1	Feature class 2	Compatible spatial relationship types
Multipoint	Point	• Contains • Intersects • Relation • Within
Multipoint	Multipoint	• Contains • Intersects • Overlaps • Relation • Within
Multipoint	Line	• Crosses • Intersects • Relation • Touches • Within
Multipoint	Area	• Crosses • Intersects • Relation • Touches • Within
Line	Point	• Contains • Intersects • Relation • Touches
Line	Multipoint	• Contains • Crosses • Intersects • Relation • Touches
Line	Line	• Contains • Crosses • Intersects • Overlaps • Relation • Touches • Within Note: With the Contains, Relation, and Within spatial relationship types, you can merge features from feature class 2 and find features from feature class 1 with a spatial relationship to the merged lines.

Feature class 1	Feature class 2	Compatible spatial relationship types
Line	Area	• Crosses • Intersects • Relation • Touches • Within Note: With the Relation and Within spatial relationship types, you can merge features from feature class 2 and find features from feature class 1 with a spatial relationship to the merged polygons.
Area	Point	• Contains • Intersects • Relation • Touches
Area	Multipoint	• Contains • Crosses • Intersects • Relation • Touches
Area	Line	• Contains • Crosses • Intersects • Relation • Touches Note: With the Contains and Relation spatial relationship types, you can merge features from feature class 2 and find features from feature class 1 with a spatial relationship to the merged lines.
Area	Area	• Contains • Intersects • Overlaps • Relation • Touches • Within Note: With the Contains, Relation, and Within spatial relationship types, you can merge features from feature class 2 and find features from feature class 1 with a spatial relationship to the merged polygons.

In addition to spatial analysis, the Compare Attributes dialog box optionally allows attributes between feature classes to be compared. For example, in the Nautical S-57 data model, sounding points that exist within depth area polygons must have depth (z) field values that are within the minimum and maximum depth (z) field values specified in the depth area polygon that contains them. SQL where clauses can be constructed on the Compare Attributes dialog box to perform attribute comparison on the features along with spatial analysis.

Intersects or Touches and Result Geometries

The Geometry on Geometry check creates result geometries if features from either the same or two different feature classes share a spatial relationship. If this check uses the Intersects or Touches spatial relationship, result geometries will be points. For example, if you configure the Geometry on Geometry check to validate two polygon feature classes with the Intersect operator, all result geometries will be points. Result points will be created where polygons in the two feature classes intersect.

Inverse Relationships

With the Geometry on Geometry check, you can also find features that do not share the spatial relationship or spatial and attribute relationship specified. In this scenario, the check finds features that share the spatial or spatial and attribute relationship defined in the check and returns the features from feature class 1 and feature class 2 that do not meet the criteria.

A simple scenario for this is rivers that intersect lakes. The check can be used to find rivers that do not intersect any lakes and lakes that are not intersected by rivers. Configuring the check to find this relationship only requires that you define the spatial relationship for the two feature classes and check the Not find features not in this relationship check box.

For example, you can find lines and polygons that do not intersect and share the same subtype code. This means that if a line intersects a polygon with a different subtype code, it will be returned as a result. Lines that do not intersect a feature at all but are of a different subtype would also be returned as results. To configure the check to find this inverse relationship you would do the following:

- Define the spatial relationship for feature class 1 and feature class 2.

- Set the attribute comparison so the subtype of data source 1 is equal to the subtype of data source 2.

- Check the Not - find features not in this relationship check box.

You can find inverse relationships using the Contains, Crosses, Intersects, Overlaps, Relation, Touches and within relationship types.

Spatial Relationship Relation

When you choose Relation as the spatial relationship type, you can compare any possible spatial intersections between two shapes based on the following three aspects:

- Interior: The entire shape, except for its boundary. All geometry types have interiors.

- Boundary: The endpoints of all linear parts for line features, or the linear outline of a polygon. Only lines and polygons have boundaries.

- Exterior: The outside area of a shape. All geometry types have exteriors.

This spatial relationship is defined using a nine-character string composed of the following characters:

- T (true): The features have interiors, boundaries, and/or exteriors that intersect.

- F (false): The features do not have interiors, boundaries, and/or exteriors that intersect.

- 0 (non-dimensional): The intersection between the interiors, boundaries, and/or exteriors of the features forms a point.

- 1 (one dimensional): The intersection between the interiors, boundaries, and/or exteriors of the features forms a line.

- 2 (two dimensional): The intersection between the interiors, boundaries, and/or exteriors of the features forms a polygon.

- *(do not check): An aspect of the relationship between the interiors, boundaries, exteriors is not checked.

The placement of the respective characters is important because it indicates what is going to be checked between the two features. For example, if the first character in the string is T, the Geometry on Geometry check looks at the interiors of features from both feature classes to see if they intersect; or if the first character in the string is 0, the check would return two line features that cross interiors at a point, but not two line features that have any congruent lengths.

The order of the characters is as follows:

Character number	Feature class 1	Feature class 2
1	Interior	Interior
2	Interior	Boundary
3	Interior	Exterior
4	Boundary	Interior
5	Boundary	Boundary
6	Boundary	Exterior
7	Exterior	Interior
8	Exterior	Boundary
9	Exterior	Exterior

Specific patterns that can be used to find specific relationships are listed in the following table:

Spatial relationship	Selection geometry	Requested geometry	String
Contains	Line	Line	TT*FFT***
Contains	Point	Line	TT*FFT***
Contains	Point	Point	T********
Contains	Line	Poly	TT*FFT***
Contains	Poly	Poly	TT*FFT***
Crosses	Line	Line	TF*FF****
Crosses	Poly	Line	TT**F****
Crosses	Line	Poly	TT**T****
Overlaps	Line	Line	TT*T*****

Spatial relationship	Selection geometry	Requested geometry	String
Overlaps	Point	Point	T*********
Overlaps	Poly	Poly	TT*T*****
Touch	Line	Line	FF*FT****
Touch	Poly	Line	FF*FT****
Touch	Line	Poly	FF*FT****
Touch	Poly	Poly	FF*FT****
Within	Line	Line	TF**F****
Within	Point	Line	T*********
Within	Point	Point	T*********
Within	Line	Poly	TF**F****
Within	Poly	Poly	TF**F****

Examples of strings that would be used in the Spatial Relationship text box are as follows:

Spatial relationship	String to use
Shares a boundary	****T****
Shares a boundary and interiors intersect	T***T****
Shares a boundary and interiors do not intersect	F***T****
Does not touch the boundary and interiors intersect	T***F****
Boundary of a polygon intersects the interior of a line along a congruent length	***1*****

DE-9IM

The "Dimensionally Extended 9-Intersection Model" (DE9IM) provides a framework for modelling the interaction of two spatial objects.

Every spatial object is characterized by the following spatial attributes:

- An interior

- A boundary

- An exterior

For polygons, these attributes are as follows:

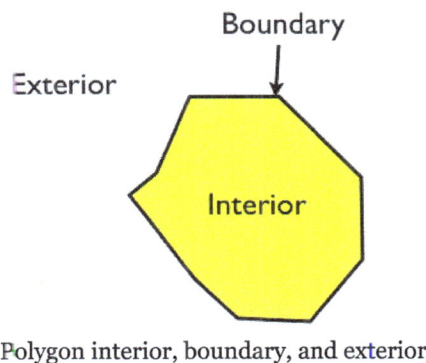

Polygon interior, boundary, and exterior

The interior is bounded by the rings, the boundary is represented by the rings themselves, and the exterior is everything else beyond the boundary. For linear features, the interior, boundary, and exterior attributes are less obvious.

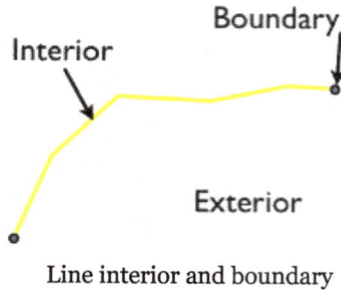

Line interior and boundary

The interior is the part of the line bounded by the ends, the boundary is represented by the ends of the linear feature, and the exterior is everything that is neither interior nor boundary.

For points, the interior is the point, the boundary is an empty set, and the exterior is everything that is not the point.

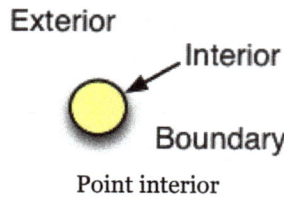

Point interior

Using these definitions of interior, boundary, and exterior attributes, the relationships between any pair of spatial features can be characterized using the dimensionality of the nine possible intersections between the interiors, boundaries and exteriors.

Modelling object interactions

For the polygons in the example above, the intersection of the interiors is a two-dimensional area, so that portion of the matrix is completed with a 2. If the boundaries intersect along a line, that portion of the matrix is completed with a 1. When the boundaries only intersect at points, which are zero-dimensional, that portion of the matrix is completed with a 0. When there is no intersection between components, the matrix is filled out with an F.

In this next example, a line string intersects a polygon:

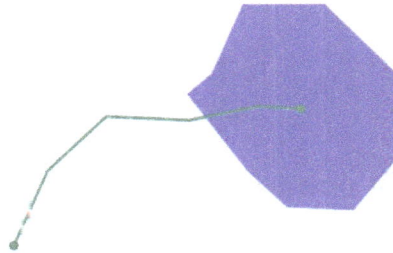

Line string intersecting a polygon

The DE9IM matrix for the interaction is represented as follows:

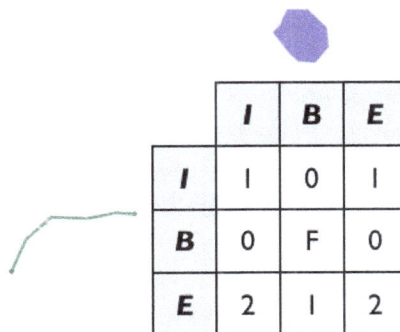

	I	*B*	*E*
I	I	0	I
B	0	F	0
E	2	I	2

DE9IM model for the intersection

Note that the boundaries of the two objects don't intersect at all (the end point of the line interacts with the interior of the polygon, not the boundary, and vice versa), so the B/B (boundary/boundary) cell is completed with an F.

DE9IM Model in PostGIS

In PostGIS, the ST_Relate function will compile a DE9IM matrix and return a string representing the DE9IM relationship between the two input geometries.

The previous example can be simplified using a simple box and line, with the same spatial relationship as the polygon and line string:

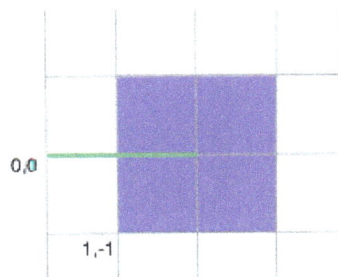

Simplified line string intersecting a polygon

With the simplified line and polygon, the line and polygon can be converted into short well-known text versions, and the DE9IM information generated in SQL as follows:

```
SELECT ST_Relate(
        'LINESTRING(0 0, 2 0)',
        'POLYGON((1 -1, 1 1, 3 1, 3 -1, 1 -1))'
        );
```

The answer, 1010F0212, is the same answer calculated above, only this time the result is returned as a nine-character string, with the first row, second row, and third row of the table appended together.

```
101

0F0

212
```

DE9IM matrices may be used in a query to find geometries with specific relationships to another geometry. For example:

CREATE TABLE lakes (id serial primary key, geom geometry);

CREATE TABLE docks (id serial primary key, good boolean, geom geometry);

INSERT INTO lakes (geom)

VALUES ('POLYGON ((100 200, 140 230, 180 310, 280 310, 390 270, 400 210, 320 140, 215 141, 150 170, 100 200))');

INSERT INTO docks (geom, good)

   VALUES

```
            ('LINESTRING (170    290,    205    272)', true),
            ('LINESTRING (120    215,    176    197)', true),
            ('LINESTRING (290    260,    340    250)', alse),
            ('LINESTRING (350    300,    400    320)', false),
            ('LINESTRING (370    230,    420    240)', false),
            ('LINESTRING (370    180,    390    160)', false);
```

The example data comprises two objects, Lakes and Docks. For this example, docks must be inside lakes and must touch the boundary of their containing lake at only one end.

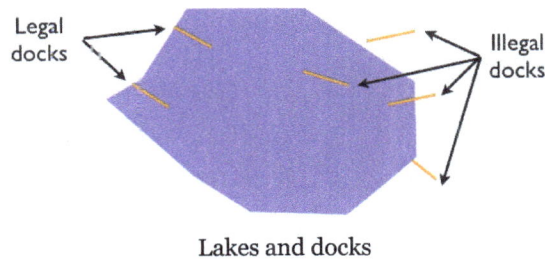

Lakes and docks

Legal docks, docks that obey the data quality rules, have the following characteristics:

- Interiors have a linear (one-dimensional) intersection with the lake interior;
- Boundaries have a point (zero-dimensional) intersection with the lake interior;
- Boundaries also have a point (zero-dimensional) intersection with the lake boundary;
- Interiors have no intersection (F) with the lake exterior.

The resulting DE9IM matrix is:

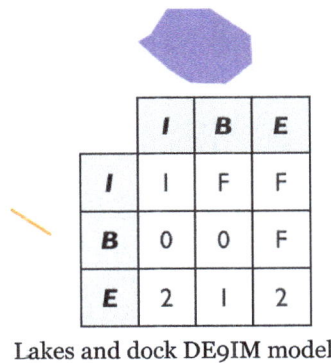

	I	B	E
I	I	F	F
B	0	0	F
E	2	I	2

Lakes and dock DE9IM model

To find all the legal docks, identify the docks that intersect lakes (a super-set of potential candidates used as the join key) and then find all the docks in that set which have the legal relate pattern.

```
SELECT docks.*
FROM docks JOIN lakes ON ST_Intersects(docks.geom, lakes.geom)
WHERE ST_Relate(docks.geom, lakes.geom, '1FF00F212');
```

This Identifies Two Valid Docks

Note the use of the three-parameter version of ST_Relate, which returns true if the pattern matches or false if it does not. For a fully defined pattern like this one, the three-parameter version is not required and a string equality operator could have been used.

However, for less rigorous pattern searches, the three-parameter allows substitution characters in the pattern string:

- "*" —Any value in this cell is acceptable
- "T"—Any non-false value (0, 1 or 2) is acceptable

So for example, one dock not included in the example illustration is a dock with a two-dimensional intersection with the lake boundary.

```
INSERT INTO docks (geom, good )

VALUES ('LINESTRING (140 230, 150 250, 210 230)',true);
```

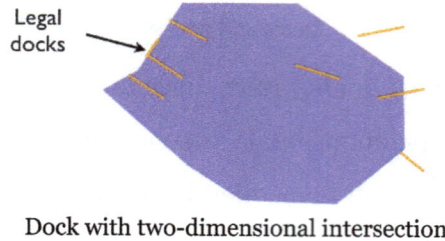

Dock with two-dimensional intersection

To include this case in the set of "legal" docks, change the relate pattern in the query. In particular, the intersection of the dock interior and lake boundary can now be either **1** (the new case) or **F** (the original case). To identify this case, use the "*" catchall in the pattern.

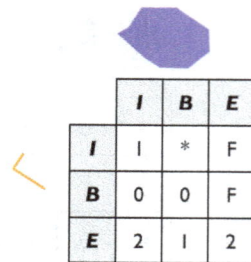

	I	B	E
I	1	*	F
B	0	0	F
E	2	1	2

Using the catchall pattern

The resulting SQL is as follows:

SELECT docks.*

FROM docks JOIN lakes ON ST_Intersects(docks.geom, lakes.geom)

WHERE ST_Relate(docks.geom, lakes.geom, '1*F00F212');

This will identify all three valid docks.

Data Quality Testing

TIGER (Topologically Integrated Geographic Encoding and Referencing) census data is quality controlled according to strict data model rules. For example, no census block should overlap any other census block.

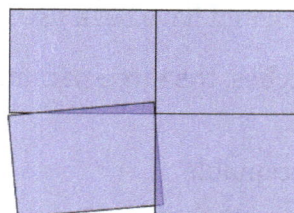

Overlapping census blocks

The following SQL command will test for any overlaps. The matrix value ('2********') represents an overlap of two interiors.

```sql
SELECT a.gid, b.gid
FROM nyc_census_blocks a, nyc_census_blocks b
WHERE ST_Intersects(a.the_geom, b.the_geom)
  AND ST_Relate(a.the_geom, b.the_geom, '2********')
  AND a.gid != b.gid
LIMIT 10;
```

This return **0**, confirming the data is clean and no overlaps were detected.

Similarly, the TIGER data model also requires all roads data to be end-noded which means intersections only occur at the ends of each street, not at the mid-points.

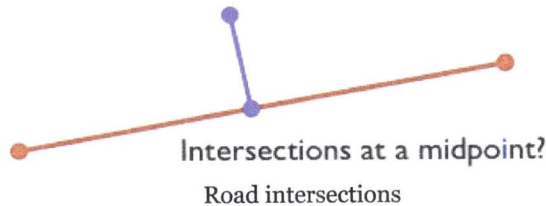

Intersections at a midpoint?

Road intersections

To test for this data model error, search for streets that intersect, using a join operation, but where the intersection between the boundaries is not zero-dimensional (the end points don't touch).

```sql
SELECT a.gid, b.gid
FROM nyc_streets a, nyc_streets b
WHERE ST_Intersects(a.the_geom, b.the_geom)
  AND NOT ST_Relate(a.the_geom, b.the_geom, '****0****')
  AND a.gid != b.gid
LIMIT 10;
```

If the result indicates the end points do not intersect, the data is not end-noded and violates the census data model rules.

Topology Analysis

Topological data analysis has been very successful in discovering information in many large and complex data sets.

One of the key messages around topological data analysis is that data has shape and the shape matters. Although it may appear to be a new message, in fact it describes something very familiar.

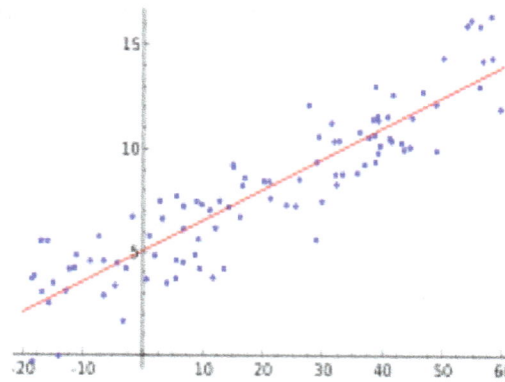

The example above is a regression line, obtained by fitting a straight line to the data points using a natural measure of fit. A straight line is certainly a shape, and in the above example, we find that a straight line fits the given data quite well. That piece of information is extremely important for a number of reasons. One is that it gives us the qualitative information that the y-variable varies directly with the x-variable (i.e. that y increases as x increases). Another is that it permits us to predict with reasonable accuracy one of the variables if we know the value of the other variable. The idea is that the shape of a line is a useful organizing principle for the data set, which permits us to extract useful information from it.

Unfortunately, the data does not always cooperate and fit along a line. Consider, for example, the data set below.

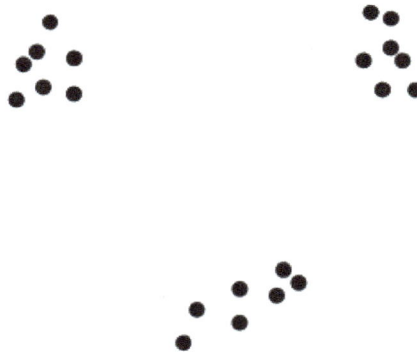

It is easy to see that no straight line faithfully represents this data.

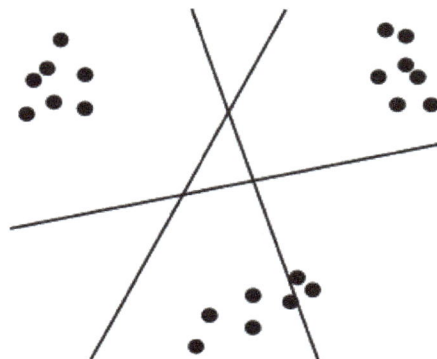

The reason is that this data set breaks into a set of three tightly concentrated clusters. One might not initially think of this as having anything to do with shape, but after a moment's reflection, we realize that the most fundamental aspect of any shape is the number of connected pieces it breaks into. So, in this case, we see that the shape of this data set is of fundamental importance, and that its shape is not that of a line.

At this point, we might think that we could now proceed by assuming that any data set is well approximated by a line, a family of clusters, or perhaps a family of lines. Here is another data set that demonstrates that this is not the case.

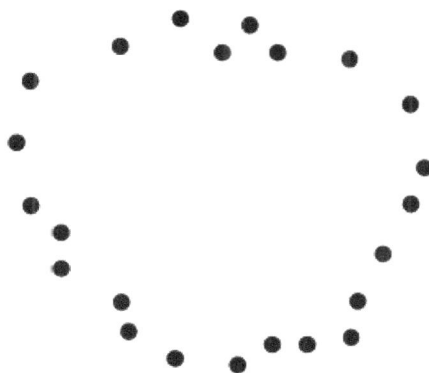

Notice that this shape also does not fit along a line, and does not break into clusters, but rather has a "loopy" behavior. This kind of structure is often associated with periodic or recurrent behavior in the data set. Here is another example.

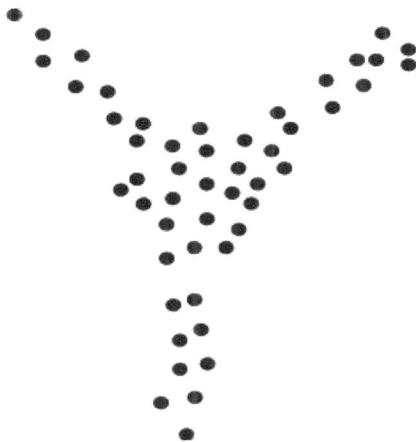

The shape is in this case that of a capital letter "Y". This is another kind of shape, which also occurs frequently. Note that it has a central core and three "flares" extending from it. This might represent a situation where the core represents the most frequently occurring behaviors and the tips of the flares represent the extreme behaviors in the data. It is clearly distinct from the three other shapes we have already discussed.

One might now say that a way to understand data would be to take each of these types, and attempt to fit a template for each to the data to determine which type one is in. This fitting process is what is done in linear regression, which is the first example above. The problem with this approach is that there are an infinite variety of different possible shapes, a large number of which occur in real data sets. All four that we have shown certainly do, but many others do as well as demonstrated in the image below.

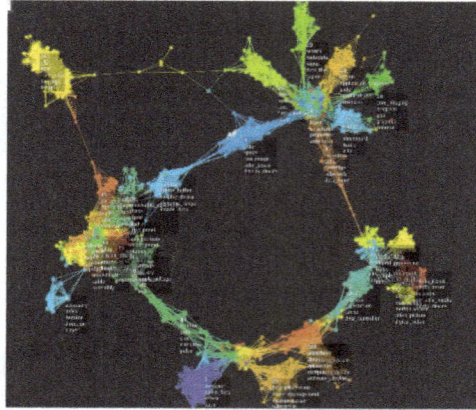

The immense variety possible among shapes suggests that we should not attempt to enumerate all the possible shapes, and create templates for each, but rather find a flexible way of representing all shapes, That is one of the problems topological data analysis deals with.

To give an idea of why Topological Data Analysis often works better than other methods of displaying data sets, such as scatterplots based on principal component analysis or multidimensional scaling, it is useful to consider another example.

This is a data set concentrated along an ellipse, where one axis is much longer than the other. If one applies principal component analysis or multidimensional scaling to this data set, the distances in the horizontal direction may be too small to recognize, and the data set will appear to be a line segment rather than an ellipse. One aspect of TDA is the construction of network representations of data sets. To see how this works for this data set, consider the first principal component projection for this data set.

The projection is visualized as moving from left to right. What is then done in the Ayasdi Core software application is to bin the data using overlapping intervals in the projected space on the right. Each bin is then clustered to produce nodes for a network, and the edges are inserted whenever two of the clusters contain a data point in common. This is possible because the bins overlap. The effect of this on a single bin is pictured as below.

We will then obtain two clusters for each bin except at the top and bottom, and a single cluster for each of those, yielding a circular network. The reason this gives more sensitivity is that the clustering only depends on the relative distances within the bins, and not on its absolute magnitude. So, as long as the horizontal distances between points are significantly larger than the vertical inter point distances, we will capture the two distinct clusters, even if the absolute magnitude of the horizontal distances is very small.

This procedure gives additional resolution as well as additional flexibility in the representation of the data. That is the power of topological data analysis.

With the introduction of sensors in everything and online systems with click by click data on all user activity, data science now touches nearly every field of study. However, the traditional techniques of data analysis have not always kept up with the exploding quantity and complexity of data since they often rely on overly simplistic assumptions. The field of topological data analysis (TDA) has attempted to fill this void by producing a collection of techniques stemming from the idea that data has shape that can be rigorously quantified in order to investigate data. This quantification takes the form of a topological signature: a representation of some aspect of the shape in a simplified form for study. As with any summary, producing a topological signature from data is a lossy process; that is some information will be lost during the creation of the summary. However, the art of using TDA is to put the data in a form that both fits into the standard TDA pipelines and where the lossy-ness of the method serves to remove high-dimensionality rather than important structure.

Data with Distance

The first hurdle to clear is to have a universal understanding for what we mean by the word "data" as this term means many different things in many different domains. Here, we will take data to mean a collection of data points with each arising from, say, different people, different students, different times and so on.

Much of TDA is based around the notion that there is an idea of proximity between these data points. So, for example, if each data point $x = \{x_1, \cdots, x_n\}$ consists of n numerical values, we have an easy definition of proximity that comes from the standard Euclidean distance: this is the generalization of the standard distance in the plane $d(x,y) = \sqrt{(x_1 - y_1)^2 + (x_2 - y_2)^2}$. Euclidean distance gives a good intuitive starting place for the requirements of a generalized distance in the mathematical sense. A distance is a function on two inputs (x, y) which satisfies the commonly used properties of the Euclidean distance, namely:

- (Positivity) $d(x, y) \geq 0$, and $d(x, y) = 0$ if and only if $x = y$. The distance between two points is a positive number, and the distance is 0 if and only if the points are the same.

- (Symmetry) $d(x, y) = d(y, x)$. The distance from one point to another is the same as going in the opposite direction.

- (Triangle inequality) $d(x, z) \leq d(x, y) + d(y, z)$. The distance between two points x and z is always no longer than taking a detour through point y.

When we have a collection of data points with a definition of a distance, we often refer to this collection as a point cloud. See, for example, the black dots of figure, which inherit a distance from being inside of the plane.

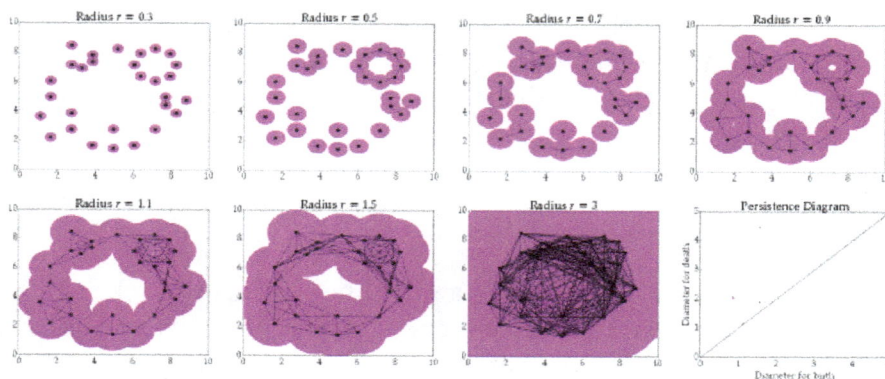

Figure: An example using persistent homology to investigate a point cloud data set by constructing the Rips Note that the Euclidean distance assumes that the inputs are numeric. There are, of course, other distances that can be defined on numeric data as well (i.e., the Minkowski distance); however, having entirely numeric data is not a requirement to define a generalized distance. One can also define many different distances when the data is categorical rather than numeric. This might simply be done by looking at matches (define the distance between data points by the number of entries which are the same), or by including a more nuanced view of the categorical entries.

Persistent Homology

The first topological signature comes from persistent homology, a powerful tool in TDA for investigating the structure of data. The persistence diagram can show a great deal of information about a given point cloud such as clustering without an expert-chosen connectivity parameter, which is usually necessary. It can also describe more complicated structure such as loops and voids that are not visible with other methods. Persistent homology has found success in the investigation of data from many different domains; these include image processing.

Simplicial Complexes

The main goal of TDA is to investigate the intrinsic shape of the data using a provided distance. However, the data as provided is nothing more than a collection of individual points, often with too many coordinates each to be fully visualizable. For instance, the point cloud in figure seems to be sampled from some sort of circular structure, but how can that structure be found or represented, particularly if the data came with, say, 73 coordinates instead of 2 as drawn? Thus, we need a structure that can be used as a proxy for the shape during our investigations.

Graphs are a commonly used structure in many data analysis applications since they can store relationships between data points. In a way, graphs encode a 1-dimensional skeleton of the data. That is, the vertices can be thought of as 0-dimensional objects, and the edges as 1-dimensional objects. But, like a skeleton, there are higher dimensional relationships that are lost when we can only see the skeleton. Think of the human forearm: if we could only see the radius and ulna bones, we would think the human arm had a gaping hole. When we can pass to understanding not only the human skeleton but the muscle, too, we realize that arm-hole is filled in and so it is not an inherent part of the body's structure. Simplicial complexes generalize the notion of graphs by allowing for 2-, 3-, and higher dimensional building blocks, called simplices.

This can be seen in the following way. Let us start building a simplicial complex with each data point representing a vertex. A vertex, or "0-dimensional simplex," consists of, obviously enough, one vertex. An edge, or "1-dimensional simplex," is defined by its two endpoint vertices. A 2-dimensional simplex is a triangle, given by its three vertices, and so on. Generally, a d-dimensional simplex is defined by $d + 1$ vertices.

A face of a simplex is one defined by a subset of its vertex set. So, the faces of an edge (given by 2 vertices) are the two endpoint vertices (and technically the edge itself). The faces of a 2-simplex (triangle given by 3 vertices) are the three vertices, the three edges, and the triangle itself. A collection of simplices is called a simplicial complex if all faces of any simplex in the collection are also in the collection. For example, a 2-simplex cannot be in the simplicial complex unless all its edges are also. An example of a simplicial complex which has 0-, 1-, 2-dimensional simplices.

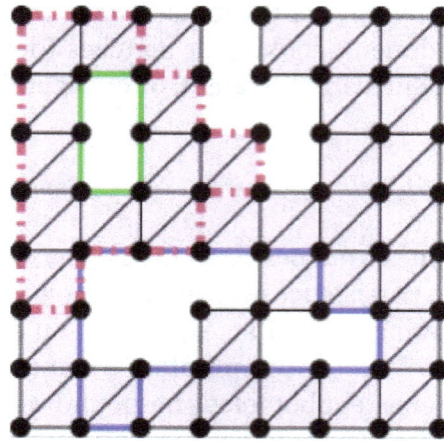

Figure: An example of a simplicial complex. The green, pink, and blue collections represent 1- dimensional homology classes, where the pink and green classes are said to be equivalent since they encompass the same hole. The rank of the 1-dimensional homology is 3 due to the three holes in the space.

The Rips Complex

The next task is to build a useful simplicial complex representing the structure of the data and which uses the original data as the vertex set. Assume we have decided on a metric for the data points and pick a number $t \geq 0$ to start. The vietoris Rips complex (sometimes called Rips complex) for parameter t is constructed in the following way. The vertex set is given by the data set itself. For each pair of points x, y in the data set, we include the edge xy if the distance between them is at most t: $d\,(x, y) \leq t$. For a higher dimensional simplex given by vertices $x_o, \cdots, x_d$, we include the simplex if the complex has all possible edges; explicitly, this means that every vertex $x_o, \cdots, x_d$ is within distance t of every other vertex in the simplex.

The Rips complex is shown in figure for several different choices of t. Note that saying that two points are within distance t of each other is the same as saying that the pair of disks of radius $r = t\,/\,2$ centred at each point touch. So in the figures, we have an edge whenever two of the disks intersect, and we assume that we have triangles and higher dimensional simplices whenever possible although these are not explicitly drawn. The data of this example appears to have come from a circular structure. We can see that there is a range of parameter values for which the Rips complex has this circular structure, namely from approximately $t = 1.8$ (radius = .9) to $t = 4.2$ (radius = 2.1). The question remains, however, how to do a good job of choosing the t parameter so that the Rips complex reflects the structure of the underlying data set. It is exactly this question that leads us use the persistence diagram as a topological signature of the data.

Persistence Context

As we have seen, the Rips complex is particularly useful for seeing structure in the data as long as the connectivity parameter t is chosen well. But, how can we do a good job of choosing t?

The answer is to not choose t, but instead to look at the continuum of possible t values looking for ranges that seem to represent something interesting in terms of structure. There are two pieces to the idea of persistent homology. Homology is a tool from classical algebraic topology that can measure certain structures of a simplicial complex. The "persistent" part comes from looking at all possible t values to see where structure appears and disappears; that is, the collection of t values for which the structure persists.

Homology is divided into different dimensions representing the dimension of the structure being measured; figure for information on homology for some commonly studied topological spaces. Here, 0-dimensional homology measures clusters; 1-dimensional homology measures loops; and 2- dimensional homology measures voids (air bubbles). There are, of course, definitions of k-dimensional homology for higher k. We lose the ability to intuitively visualize the structures we are capturing for higher k, thus these have not yet been commonly used in applications. Figure gives the so-called Betti number β_k for different example spaces. The Betti number is the rank of the k-dimensional homology group; in particular, it counts the number of structures seen in that dimension. For the purposes of concreteness, we will discuss 1-dimensional homology and persistent homology here, and defer for more rigorous and thorough introductions to general homology and persistent homology.

	β_0	β_1	β_2	β_3
.	1	.	.	.
◯	1	1	.	.
(sphere)	1	.	1	.
(torus)	1	2	1	.
(Klein bottle)	1	2	1	.

Betti numbers β_k for different topological spaces: a point, a circle, a sphere, a torus (donut), and a Klein bottle. The kth Betti number gives the rank of the k-dimensional homology group, thus measures different properties of the space in each dimension. For example, $k = 0$ measures connectivity, $k = 1$ measures loops, and $k = 2$ measures voids.

Structures in homology are given by "classes." A class in 1-dimensional homology is represented by a collection of 1-simplices (edges) that have an even number of edges touching each vertex. For example, a collection of edges that form a closed loop, as in the example of figure, satisfies this requirement. The reason for using the word "represented" is that different closed loops can represent the same class. Essentially, for two loops to represent the same class, they must encircle the same hole, such as the dashed pink and solid green loops in figure. Working with homology means that rather than studying the incredibly large collection of all such loops, we can divide them into groups where all elements of a group represent the same structure in the space.

While homology measures the structure of a single, stagnant space, persistent homology watches how this structure changes as the space changes. Consider the example in figure. In this case, we have a point cloud and can build the Rips complex for several different choices of the parameter t. As t grows, more and more edges and higher dimensional simplices are added. So, we can choose a representative for a 1-dimensional homology class at one t, and see if it still represents a class at a larger t. The t for which a class is first seen is called the birth diameter, and the t for which a class is no longer different from the previously seen classes is called the death diameter. In the example of figure, we see a 1- dimensional class (a loop) is born when the radius gets to $r = 0.5$ ($t = 1$); however, this fills in by the time $r = 1$ ($t = 2$). On the other hand, there is a large loop born when $r = 0.9$ ($t = 1.8$), which does not die until $r = 2.1$ ($t = 4.2$).

For each class, we have a pair of numbers (a, b) for the birth diameter and death diameter. These pairs of numbers are drawn as a point in the persistence diagram. For example, there are points at (1,2) and (1.8,4.2) representing those two classes discussed in the previous paragraph. Note that if a class dies very soon after it is born (i.e., if it has a short lifetime), it will be represented by a point close to the diagonal. If the class has a long lifetime, then it will be represented as a point far from the diagonal. In many applications, the existence of a few points far from the diagonal relative to the rest of the points in the persistence diagram can be taken to mean that these classes come from the inherent structure of the data, while the rest of the points are artifacts of noise.

Of great use in the case of data with noise is the existence of a distance on the persistence diagrams themselves. This way, we have a computable measure of just how similar two persistence diagrams are. Take, for example, the two point clouds in the left of figure. The red circle point cloud is a noisy version of the black square point cloud. The persistence diagrams for the two point clouds are drawn overlaid in the right of figure. While the red diagram has quite a few more points due to noise, both diagrams have a single point far from the diagonal, thus they represent similar underlying structures.

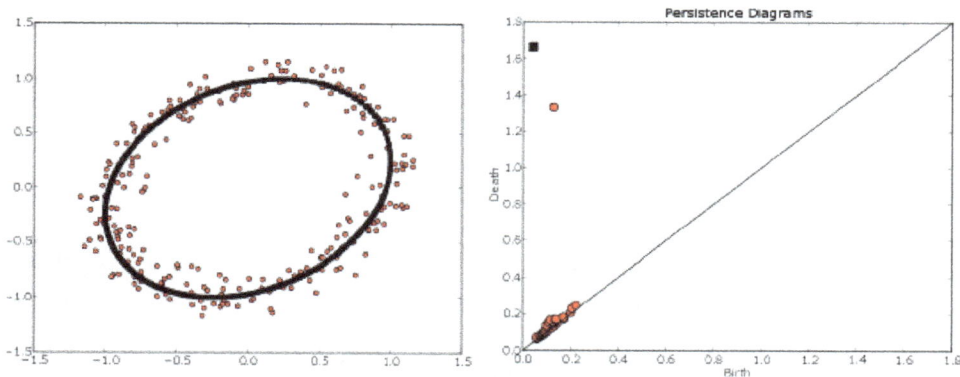

Figure: Two example point clouds are overlaid at left, and their persistence diagrams are overlaid at right. Notice that the point clouds are close in some sense. The fact that the persistence diagrams are also close is a result of the stability theorem for persistence.

Two metrics are commonly used to measure the similarity of these objects: the bottleneck and Wasserstein distances. Each works by matching points of one diagram with points of another diagram while allowing the match to be done with the diagonal if necessary. Bottleneck distance is the maximum distance between any pair of points, and thus gives a measure for the most work that must be done to push one diagram into the configuration of the other. Wasserstein distance sums powers of the distances between the pairs; unlike the bottleneck distance, it takes all of the points into account, including the noisy diagonal points. The former is better for a simple test of proximity of diagrams; the later is better when the noisy classes on the diagonal hold useful information about the data.

Arguably the most important theorem for justifying the use of persistent homology as a tool for data analysis is the stability theorem. It says that the distance between two persistence diagrams (using bottleneck distance) cannot be larger than the distance between the two data sets (using the so-called Hasudorff distance) used to obtain them. When we view this result in the context of data, it says that even if we are given a data set infected by a little noise, the persistence diagram obtained from this data is approximately correct because it is close to the diagram we would have from the noise-free data.

4 MAPPER

Another very powerful signature arising from TDA is mapper. The idea is to represent the 1-dimensional structure of a data set using a graph. Unlike the persistence diagram, data points have associated locations to points in the mapper graph. Since these graphs are much easier to visualize than the possibly high dimensional data used to construct it, mapper is an excellent tool for investigation and visualization of the structure of a data set. It has been used extensively in data analysis, particularly in the biology and health domains.

Like with the persistence diagrams, we start with a point cloud and a choice of distance on the points. However, unlike the persistence diagram, we have a few more choices to make along the way in order to construct the mapper graph.

The main additional piece of data required for the mapper graph construction is called a filter function. This filter function is simply an assignment of a real number for each data point and is used to spread out the data. Some examples include using one or a combination of the coordinates for each point, eccentricity, or density.

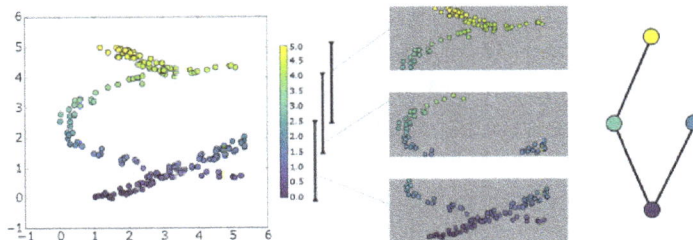

Figure a: A small example of mapper. The filter function for the given data is given by the y-coordinate. A cover is chosen for the function values, and then the point cloud is clustered within each portion defined by the cover. The resulting graph at right is the mapper graph for the chosen parameters.

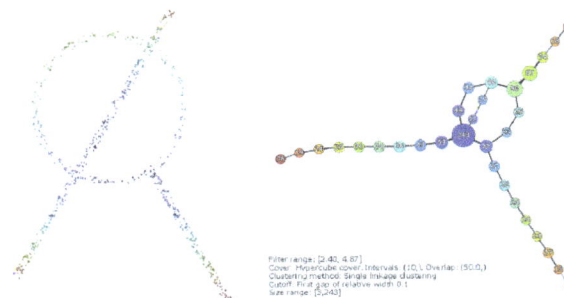

Figure b: An example point cloud with 700 points is drawn at top left. The eccentricity filter function is shown by the colouring of the points. The mapper graph for this example is shown at the top right. Note how the structure of the mapper graph reflects the intrinsic structure of the point cloud.

Then, we decide on a cover of the filter function values. A cover for the interval $[a, b]$ is a collection of overlapping sets such that each number between a and b is included in at least one set. Finally, we consider the subset of points with function values in a single set from the cover, and then cluster the points. Each cluster becomes a node in the mapper graph as in the far right of figure. Edges are included based on the intersection information from overlapping sets in the cover.

The important thing to note about the mapper graph is that each node represents a subset of the data points. This makes it a particularly useful signature for data investigation since it can separate data points with different properties even if standard clustering cannot differentiate them. For instance, the example of figure would be seen as a single cluster, but the mapper graph allows us to differentiate between points on the circle and points on the different flares.

Geography Markup Language

GML or Geography Markup Language is an XML based encoding standard for geographic information developed by the OpenGIS Consortium (OGC).

Geography, Graphics and Maps

Before we look at GML itself, it is important that we draw some clear distinctions between geographic data (which is encoded in GML) and graphic interpretations of that data as might appear on a map or other form of visualization. Geographic data is concerned with a representation of the world in spatial terms that is independent of any particular visualization of that data. When we talk about geographic data we trying to capture information about the properties and geometry of the objects which populate the world about us. How we symbolize these on a map, what colors or line weights we use is something quite different. Just as XML is now helping the Web to clearly separate content from presentation, GML will do the same in the world of geography.

GML is concerned with the representation of the geographic data content. Of course we can also use GML to make maps. This might be accomplished by developing a rendering tool to interpret GML data, however, this would go against the GML approach to standardization, and to the separation of content and presentation. To make a map from GML we need only to style the GML elements into a form which can be interpreted for graphical display in a web browser. Potential graphical display formats include W3C Scalable Vector Graphics (SVG), the Microsoft Vector Markup Language (VML), and the X3D. A map styler is thus used to locate GML elements and interpret them using particular graphical styles.

GML is Text

Like any XML encoding, GML represents geographic information in the form of text. While a short while ago this might have been considered verboten in the world of spatial information systems, the idea is now gaining a lot of momentum. Text has a certain simplicity and visibility on its side. It is easy to inspect and easy to change. Add XML and it can also be controlled.

Text formats for geometry and geography have been employed before. The pioneering work of the Province of British Columbia with its SAIF format is just one such example. In the Province of British Columbia, more than 7000 files of 1:20,000 scale data including topography, planimetry (hydrography, buildings, roads etc.) and toponymy are available in the SAIF format. The Province has shown that text formats are practical and easy to use. Another example of the use of text for complex geometric data sets is that of VRML (Vector Markup Language). Large and complex VRML models have been built and navigated over the Web all using text based encoding. Interestingly enough the VRML geometry and behaviour are themselves now being recast in XML through the efforts of the X3D Working Group.

GML Encodes Feature Geometry and Properties

GML is based on the abstract model of geography developed by the OGC. This describes the world in terms of geographic entities called features. Essentially a feature is nothing more than a list of properties and geometries. Properties have the usual name, type, value description. Geometries are composed of basic geometry building blocks such as points, lines, curves, surfaces and polygons. For simplicity, the initial GML specification is restricted to 2D geometry, however, extensions will appear shortly which will handle 2 1/2 and 3D geometry, as well as topological relationships between features.

GML encoding already allows for quite complex features. A feature can for example be composed of other features. A single feature like an airport might thus be composed of other features such as taxi ways, runways, hangers and air terminals. The geometry of a geographic feature can also be composed of many geometry elements. A geometrically complex feature can thus consist of a mix of geometry types including points, line strings and polygons.

To encode the geometry of a feature like a building we simply write:

```
<Feature  fid="142" featureType="school"  Description="A middle school">

    < Polygon name="extent" srsName="epsg:27354">

        < LineString name="extent" srsName="epsg:27354">

            < CData>

                491888.999999459,5458045.99963358 491904.999999458,5458044.99963358

                491908.999999462,5458064.99963358 491924.999999461,5458064.99963358

                491925.999999462,5458079.99963359 491977.999999466,5458120.9996336

                491953.999999466,5458017.99963357 </CData>
```

```
  < /LineString>

 < /Polygon>

</Feature>
```

Note that this has no properties (other than the geometry). These we can readily add and the building would look something like:

```
<Feature  fid="142" featureType="school" >

 <Description>Balmoral Middle School</Description>>

 <Property Name="NumFloors" type="Integer" value="3"/>

 <Property Name="NumStudents" type="Integer" value="987"/>

  < Polygon  name="extent" srsName="epsg:27354">

   < LineString  name="extent" srsName="epsg:27354">

    < CData>

      491888.999999459,5458045.99963358 491904.999999458,5458044.99963358

      491908.999999462,5458064.99963358 491924.999999461,5458064.99963358

      491925.999999462,5458079.99963359 491977.999999466,5458120.9996336

      491953.999999466,5458017.99963357 </CData>

  < /LineString>

 </Polygon>

</Feature>
```

GML Encodes Spatial Reference Systems

An essential component of a geographic system is a means of referencing the geographic features to the earth's surface or to some structure related to the earth's surface. The current version of GML incorporates an earth based spatial reference system which is extensible and which incorporates the main projection and geocentric reference frames in use today. In addition the encoding scheme allows for user defined units and reference system parameters. Future versions of GML will likely provide even more flexible encodings in order to handle local coordinate systems such as used for mile logging etc.

Why encode a spatial reference system? Why not just provide a unique name and be done with it? In many cases such an approach does suffice and GML does not require that the sender of geographic data also send an encoding of the reference system to which the data's coordinate values are referenced. There are cases, however, where such information is very valuable, and include:

- Client validation of a server specified Spatial Reference System. Client can request the SRS description (an XML document) and compare it to its own specifications or show it to a user for verification.

- Client display of a server specified Spatial Reference System.

- Use by a Coordinate Transformation Service to validate an input data sources Spatial Reference System.

- A Coordinate Transformation Service can compare the SRS description with its own specifications to see if the SRS is consistent with the selected transformation.

- To control automated coordinate transformation by supplying input and output reference system names and argument values.

With the GML encoding for spatial references, it is possible to create a web site which stores any number of spatial reference system definitions.

GML Feature Collections

The XML 1.0 Recommendation from the W3C is based on the notion of a document. The current version of GML is based on XML 1.0, and uses a Feature Collection as the basis of its document. A Feature Collection is a collection of GML Features together with an Envelope (which bounds the set of Features), a collection of Properties that apply to the Feature Collection and an optional list of Spatial Reference System Definitions. A Feature Collection can also contain other Feature Collections, provided that the Envelope of the bounding Feature Collection bounds the Envelopes of all of the contained Feature Collections.

When a request is made for GML data from a GML server, data is always returned in Feature Collections. There is no limit in the GML RFC on the number of features which can be contained in a Feature Collection. Because Feature Collections can contain other Feature Collections it is a relatively simple procedure to "glue together" Feature Collections received from a server into still larger collections.

GML - More than a Data Transport

While GML is an effective means for transporting geographic information from one place to another we expect that it will also become an important means of storing geographic information as well. The key element here is X-Link and X-Pointer. While these two specifications lag in the development and implementation area they hold great promise for building complex and distributed geographic data sets. Geographic data is, well, geographic. It is naturally distributed over the face of the earth. Interest in data about Flin Flon, Saskatchewan is much higher near Flin Flon than it would be in Pasadena, California. At the same time there are applications which need to reach out and obtain data on a global basis for large scale analysis or because of interest in a narrow vertical domain. Applications of the later sort also abound in a diverse collection of fields from environmental protection to mining, highway construction, and disaster management. How nice it would be if data could be developed on the local scale and readily integrated to the regional and the global scale?

In most jurisdictionsn geographic data is collected by particular agencies for a particular purpose. Forest bureaus collect information on the disposition of trees (tree diameters, site conditions, growth rates) for the effective management of commercial forests. Environmental departments collect information on the distribution of animals and animal habitat. Development interests maintain information on demographics and existing features in the built environment. Real world problems seldom, however, respect the parochial boundaries of departments, ministries and bureaus. How nice it would be if data developed for one purpose could be readily integrated with data developed for another?

We believe that GML as a storage format, combined with X-Link and X-Pointer will provide some useful contributions to these problems.

Technologies on which GML Depends

GML is based on XML, while sometimes talked about as a replacement for HTML, is best thought of as a language for data description. More correctly, XML is a language for expressing data description languages. XML is, however, not a programming language. There are no mechanisms in XML to express behaviour or to perform computations. That is left for other languages such as Java and C++.

XML Version 1.0

XML 1.0 provides a means of describing (marking up) data using user defined tags. Each segment of an XML document is bounded by starting and end tags. This looks as follows:

< Feature>

.... more XML descriptions ...

....

< /Feature>

The valid tag names are determined by the Document Type Definition. Which tags can appear enclosed within an opening and closing tag pair is also determined by the DTD.

XML tags can also have attributes associated with them. These are also constrained by the DTD in name and in some cases in terms of the values that the attributes can assume.

XML is typically read by an XML parser. All XML parsers check that the data is well formed so that data corruption (e.g. missing closing tag) cannot pass undetected. Many XML parsers are also validating, meaning that they check that the document conforms to the associated DTD.

Using XML is it is comparatively easy to generate and validate complex hierarchical data structures. Such structures are common in geographic applications.

XSL and XSLT (Transforming the WWW)

The original focus of XML was to provide a means of describing data separate from its presentation, especially in the context of the World Wide Web. XML Version 1.0 deals with the description of data. A companion technology, called XSL was to deal with the presentation side. Overtime it has become apparent that XSL is actually two different technologies. One, now called XSLT (the T stands for Transformation), is focused on the transformation of XML. The other technology is concerned with the actual formatting of text or images and is referred to in terms of format objects or flow objects. In our discussions we are only concerned with XSLT. Since many tools (e.g. MS IE 5.0) were developed before the XSLT label had stuck, XSL is still often used when only XSLT is intended. We will follow that practice.

If you follow xml.com, you may recall a great deal of discussion about the merits of XSL. The XSLT clarification has helped to dampen this discussion some-what however, there is still a great deal of skepticism regarding the utility and the need for XSL in some sectors of the XML community. We stand on the opposite side of the issue. We believe that it is the transformational character of XML that is most important, and XSL (XSLT) provides a clean declarative means for expressing these transformations. In our view XSLT is as essential to GML as XML itself.

XSL is a fairly simple language. It provides a powerful syntax for expressing pattern matching and replacement. It is declarative. You can easily read what the XSLT says to do. You do not get to see how it is accomplished. Using its companion specifications (X-Path and XQL) you can specify some very powerful queries on an XML document. Furthermore XSLT incorporates the ability to call functions in another programming language such as VBScript or Java through the use of Extension Functions. This means that XSL can be used to do the querying and selection, and then call out to Java or another language to perform needed computation or string manipulation. For simple tasks, XSLT provides built in string handling and arithmetic capabilities.

SVG, VML and X3D - Vector Graphics for the Web

XML has made it's presence felt in many different quarters, not the least of which is vector graphics. Several XML based specifications for describing vector graphic elements have been developed, including Scalable Vector Graphics (SVG), Microsoft's Vector Markup Language (VML), and X3D, the XML incarnation of the syntax and behaviour of VRML (Virtual Reality Markup Language). These specifications are in many ways similar to GML, but have a very different objective. Each has a means of describing geometry. The graphical specifications, however, are focused on appearance and hence include properties and elements for colors, line weights and transparency to name but a few aspects. To view an SVG, VML or X3D data file, it is necessary to have a suitable graphical data viewer. In the case of VML this is built into IE 5.0 (and nowhere else). In the case of SVG, Adobe is developing a series of plug-ins for Internet Explorer and Netscape Communicator as well as Adobe Illustrator, while IBM and several other companies, are, or have already developed SVG viewers or supporting graphics libraries. Several all Java SVG viewers are available or under development.

To draw a map from GML data you need to transform the GML into one of the graphical vector data formats such as SVG, VML or VRML. This means to associate a graphical "style" (e.g. symbol, colour, texture) with each type of GML feature or feature instance.

Figure illustrates the drawing of map using an XSLT style sheet on a suitable mapping client.

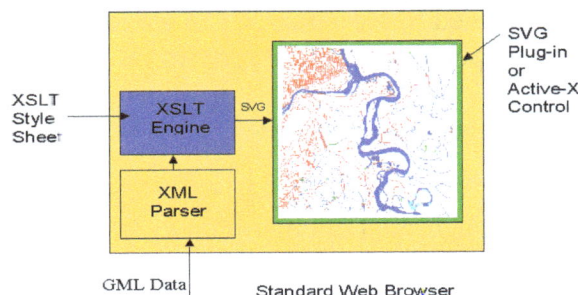

Figure: Making a Map with XSLT and SVG

X-Link and X-Pointer - Linking One place to Another

With current HTML technology it is possible to build linked geographic data sets. One can readily build image maps which are linked to other image maps. The HTML linking mechanism has, however, many limitations, and as a result it is not practical to build large complex distributed data sets as occur in real world systems. The most significant limitation is that an HTML link is effectively hard coded in both the source (<a href = ... >) and target (anchor) documents a fact which would any significant system both fragile and impossible to scale. X-Link gets around these problems by allowing "out of line" links. In an out of line link, the source points only to a link database and it is the link database that provides the pointer to specific XML elements in the target document. The link is thus not hard coded in either document. This is of great importance in relation to GML as it makes it possible to build scalable, distributed geographic data sets. Even more importantly, the X-Link and X-Pointer make it possible to build application specific indexes for dates. Need to have a group of buildings organized by street address Want to create a farm plot index based on crop type? With X-Link and X-Pointer, these and many other indexing schemes can be readily constructed, and all without altering the source data itself.

Significance of GML

Why introduce GML at all? There are already a host of encoding standards for geographic information including COGIF, MDIFF, SAIF, DLG, SDTS to name only a few. What is so different about GML? In some ways nothing, GML is a simple text based encoding of geographic features. Some of these other formats are not text based, however, some of them (e.g. SAIF) certainty are. GML is based on a common model of geography (OGC Abstract Specification) which has been developed and agreed to by the vast majority of all GIS vendors in the world. More importantly, however, GML is based on XML. Why should this matter? There are several reasons why XML is important. To begin with XML provides a method to verify data integrity. Secondly, any XML document can be read and edited using a simple text editor. Nothing more than MS Notepad is required to view or change an XML document. Thirdly, since there are an increasing number of XML languages, it will be more and easier to integrate GML data with non-spatial data. Even in the case of non-XML non-spatial data this is the case. Perhaps, most importantly, XML is easy to transform. Using XSLT or almost any other programming language (VB, VBScript, Java, C++, Java script) we can readily transform XML from one form to another. A single mechanism can thus be employed for a host of transformations from data visualization to coordinate transforms, spatial queries, and geo-spatial generalization.

GML rests securely on a widely adopted public standard, that of XML. This ensures that GML data can be viewed, edited and transformed by a wide variety of commercial and free ware tools. For the first time we can truly talk about open geographic information.

Automated Verification of Data Integrity

One of the important features of XML is the ability to verify data integrity. In the XML 1.0 Recommendation this is achieved through the Document Type Definition (DTD). The DTD specifies the structure of an XML document in a such a way that a validating parser can verify that a given document instance complies with this DTD. GML is specified by such a DTD. Future versions of GML will also be supported by XML Schema, a more flexible integrity mechanism than the DTD that should become a W3C Recommendation early in 2000.

Using the GML DTD, servers and clients can readily verify that the data they are to send or receive complies with the specification. Furthermore this can be accomplished with a variety of parsing tools by at least a have a dozen different vendors on a wide variety of operating systems, databases, application servers and browsers.

GML can be Read by Public Tools

As we have already noted, GML is text and one need have nothing more than a simple text editor to read it. GML, however, is structured, and any of a variety of XML editors can be employed to display that structure. This makes viewing and navigating GML data very easy as shown in figure below.

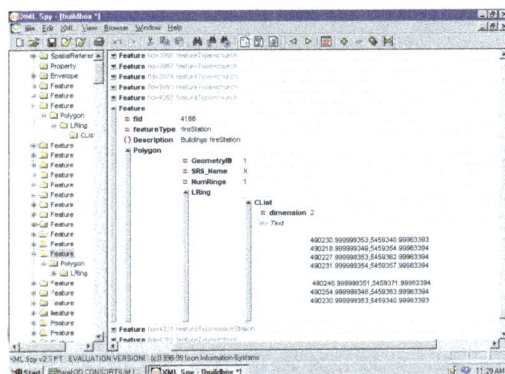

Figure: Sample GML File Viewed in XML Spy

GML can be Easily Edited

Using the many XML editors it is also very easy to edit GML data. Want to add a new feature property or change a property value? Need to adjust feature geometry. These are easily accomplished with a standard XML editor. Unlike many other texts based formats however there is no way you can corrupt the data using an XML editor. The editor can be made to ensure that any data which is created or modified complies with the DTD.

It is also not difficult to create a graphical editor for GML and such products are expected to appear on the market within the coming year. Again the GML DTD can be used to ensure data integrity. Note that when one edits GML graphically an intermediate graphic representation is required (perhaps SVG) which is then used to define the geometry of the associated GML feature.

GML can Readily Integrate with Non-spatial Data

Binary data structures are typically very difficult to integrate with one another. A classic example is that of associating a text document, or a parameter list, with a separately developed and maintained spatial database of parcels or land tenure boundaries. With a binary data structure one must understand the file structure or database schema and be able to modify it. In many legacy systems using flat files the data structure cannot be modified without breaking the applications which rely on the existing data structure. With GML it is comparatively easy to provide links to other XML data elements and this will dramatically improve with the introduction of X-Link and X-Pointer. Even links to non-XML elements can be readily handled using the well-established URI syntax.

GML is Transformable

The most important aspect of XML in our view is its transformability. It is quite easy to write a transformation which carries XML data relative to one DTD to XML relative to another. This is exactly what we do when we generate an SVG graphical element stream from a GML data file. Such transformations can be accomplished using a variety of mechanisms including XSLT, Java, Java script and C++ to name only a few. XSLT in our view is of particular interest. With XSLT it is very easy to write a style sheet which locates and transforms GML elements into other XML elements. Where XSLT is not up to the task, one can readily incorporate XSLT extension functions written in Java or VB (the exact languages supported depends on the implementation) to perform tasks such as string manipulation or mathematical computation. XSLT can also make use of powerful searching syntax (X Path/XQL) so as to retrieve elements that satisfy complex boolean expressions on the elements and their attributes. Using these techniques an XSLT style sheet can perform a wide variety of querying, analysis and transformation functions. Consider the following examples:

Using XSLT with suitable extension functions we can extract spatial elements which satisfy various spatial and attribute queries. Galdos Systems Inc. will be providing just such a set of spatial extension functions in the near future on the Geo Java site. Using these functions it will be straightforward to write a spatial query that extracts features of a given type which lie within a specified region or which intersect a particular feature.

Change the XSLT style sheet and we can accomplish a totally different function. We can for example write a style sheet that performs coordinate transformation. This immediately provides us with a coordinate transformation service. Locate GML data in one part of the world in reference system X and simply pass its URI to the service and specify the target reference system, and presto you will have GML in the new frame of reference.

Change the XSLT style sheet and we can accomplish yet another function. We can for example generate an SVG, VML or X3D map on the server. Select different style sheets for different viewing devices or different types of maps.

The transformability of GML also means that we can readily construct application specific indexes or at least we will be able to once X-Link and X-Pointer implementations start to move toward reality. It might have a huge impact on the utility of GML data sets.

GML can Transport Behaviour

XML is a language for describing data description languages. GML does not itself encode behaviour. GML can, however, be used in conjunction with languages like Java or C++ to in effect transport geographic behaviour from one place to another. This can be done using a simple object factory which instantiates objects based on received GML data, mapping the GML element names into object classes. In the Java case this would mean mapping the GML elements into Java classes as listed in the OGC Java Simple Features RFC. This "re-hydration" of the GML data then creates Java objects which have the OGC interfaces for Simple Features (of course we did not transport the interfaces). GML and Java (or COM or CORBA) Simple Features can thus get along very well with one another. In many applications one only needs the behaviour for a small number of the elements. With this approach one might receive 10,000 GML elements but only need to construct a hundred or so Java objects on an as needed basis.

Geographic Resources Analysis Support System

Bringing Advanced Geospatial Technologies

GRASS GIS (Geographic Resources Analysis Support System) is one of the oldest public domain GIS software in existence. It's more than 30 years old.

The US Army Corps of Engineers developed it as a land management and environmental planning tool.

Now, academics, government agencies (NASA, NOAA, USDA and USGS) and GIS practitioners use this open source software because its code can be inspected and tailored to their needs. But what about for the every day GIS user?

Traditionally, users have all their data stored all over the place their local drives, network drives and external ports.

While ArcGIS and QGIS will project your data on the fly, GRASS GIS insists that data in different projections be placed into separate folder locations.

"All data in one Location, is in the same coordinate reference system (projection). One Location can be one project. Location contains Map sets."

For quality reasons, GRASS GIS handles one projection per folder location. This is mostly because a lot of users out there tend to ignore their coordinate systems and simply assumes their data lines up.

Case in point, you will have to be more cautious with your data's coordinate system in a GRASS Workspace File (*.gxw).

All the Firepower You Need with Sophisticated Tools

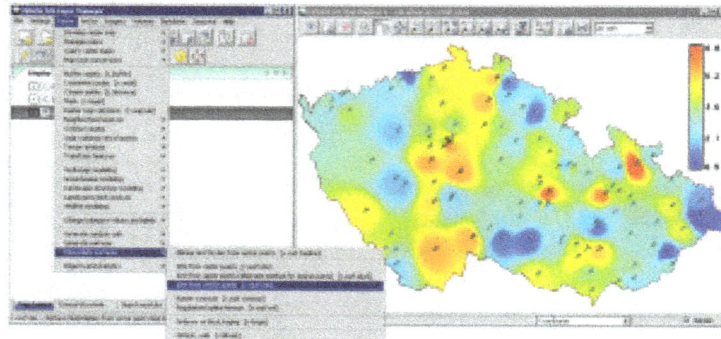

Without a doubt, this software has some serious firepower with over 400 raster and vector manipulation tools.

It has so much in its arsenal that QGIS and uDig have adopted it as a base analysis set of tools of their own.

Raster, vector, imagery, 3D raster and temporal tools are at your disposal. Some of our stand-outs included the satellite imagery tools (atmospheric correction, DN conversion and vegetation indices).

Also notable are the modelling tools for groundwater, wildfire, hydrology and landscape patch analysis.

GRASS GIS have the sophisticated analysis tools for almost any type of analysis.

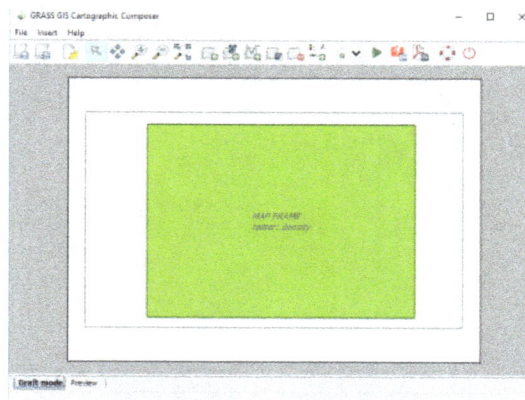

GRASS GIS is probably not suited for 99% of your mapping you need in everyday life.

The GRASS GIS Cartographic Composer is meant to interactively design and generate hard copy maps.

Other Exceptional Standouts

The 3D capabilities are surprisingly impressive. It's integrated within the map display and all you have to do is select the 3D drop down. Choose the lighting, perspective, tilt it's really quite good.

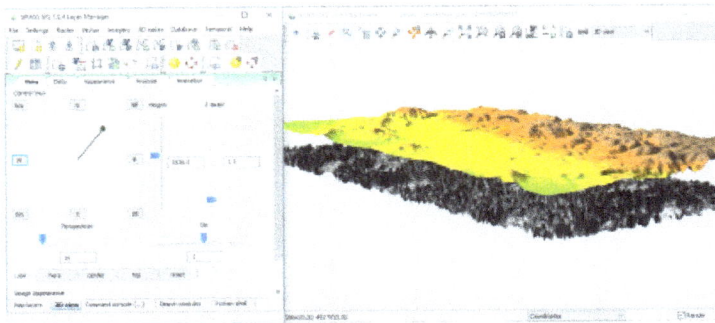

GRASS GIS can also set up vector networks with some topology maintenance. The network analysis tool includes shortest path, moving-salesman, maximum flow and center allocation.

Data interoperability is decent. Multiple data input formats are available, including MySQL, .DBF, Post GIS, and SQLite. You can't compare it to Feature Manipulation Engine (FME). Using the *Common Import Formats* OGR can be used for reading, writing and transforming vector data.

Lastly, the documentation is superb. The GRASS manual details the use of modules distributed with Geographic Resources Analysis Support System (GRASS).

Expect the Unexpected for User Interface

Some of the things you'd expect that are in standard GIS software, just aren't there. The initial settings and user interface:

- Right-clicking a layer in the table of contents to zoom to feature does not exist;
- A toolbox that separates the tools for organization is hidden in the *'Search Modules'* tab;
- You have a command line window running in the background the whole time the program is open.

GRASS GIS is set up differently than other GIS software.

The original user interface of GRASS was in command line only. It's come a long way to provide a GUI to the public – especially in the last decade or so.

References

- Geodatabase-topology-rules-and-topology-error-fixes: help.arcgis.com, Retrieved 25 May 2018
- Understand-spatial-relations: edndoc.esri.com, Retrieved 31 March 2018
- Types-of-spatial-relationships-that-can-be-validated, data-reviewer, extensions: desktop.arcgis.com,
- Retrieved 23 March 2018
- Why-topological-data-analysis-works: ayasdi.com, Retrieved 13 May 2018
- Grass-gis-geographic-resources-analysis-support-system: gisgeography.com, Retrieved 30 April 2018

5

Geostatistics

Geostatistics is a sub-field of statistics. It studies spatial or spatiotemporal data. It has applications in diverse areas of petroleum geology, hydrology, meteorology, geochemistry, geometallurgy, etc. Some of the vital tools and techniques used in geostatistics such as variogram, kriging, geodemographic segmentation and geotargeting, etc. have been extensively discussed in this chapter.

Geostatistics is a class of statistics used to analyze and predict the values associated with spatial or spatiotemporal phenomena. It incorporates the spatial (and in some cases temporal) coordinates of the data within the analyses. Many geostatistical tools were originally developed as a practical means to describe spatial patterns and interpolate values for locations where samples were not taken. Those tools and methods have since evolved to not only provide interpolated values, but also measures of uncertainty for those values. The measurement of uncertainty is critical to informed decision making, as it provides information on the possible values (outcomes) for each location rather than just one interpolated value. Geostatistical analysis has also evolved from uni to multivariate and offers mechanisms to incorporate secondary datasets that complement a (possibly sparse) primary variable of interest, thus allowing the construction of more accurate interpolation and uncertainty models.

Geostatistics is widely used in many areas of science and engineering, for example:

- The mining industry uses geostatistics for several aspects of a project: initially to quantify mineral resources and evaluate the project's economic feasibility, then on a daily basis in order to decide which material is routed to the plant and which is waste, using updated information as it becomes available.

- In the environmental sciences, geostatistics is used to estimate pollutant levels in order to decide if they pose a threat to environmental or human health and warrant remediation.

- Relatively new applications in the field of soil science focus on mapping soil nutrient levels (nitrogen, phosphorus, potassium, and so on) and other indicators (such as electrical conductivity) in order to study their relationships to crop yield and prescribe precise amounts of fertilizer for each location in the field.

- Meteorological applications include prediction of temperatures, rainfall, and associated variables (such as acid rain).

- Most recently, there have been several applications of geostatistics in the area of public health, for example, the prediction of environmental contaminant levels and their relation to the incidence rates of cancer.

In all of these examples, the general context is that there is some phenomenon of interest occurring in the landscape (the level of contamination of soil, water, or air by a pollutant; the content of gold or some other metal in a mine; and so forth). Exhaustive studies are expensive and time consuming, so the phenomenon is usually characterized by taking samples at different locations. Geostatistics is then used to produce predictions (and related measures of uncertainty of the predictions) for the un sampled locations.

Here, a generalized workflow for geostatistical studies is presented, and the main steps are explained. As mentioned previously, geostatistics is a class of statistics used to analyze and predict the values associated with spatial or spatiotemporal phenomena. Geostatistical Analyst provides a set of tools that allow models that use spatial coordinates to be constructed. These models can be applied to a wide variety of scenarios and are typically used to generate predictions for un-sampled locations, as well as measures of uncertainty for those predictions.

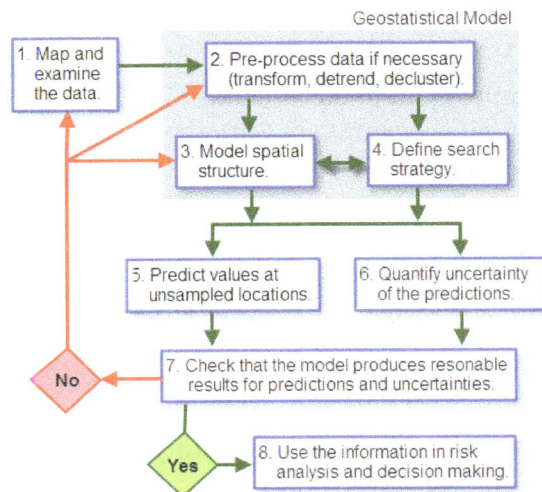

Geostatistical Model

1. Map and examine the data.
2. Pre-process data if necessary (transform, detrend, decluster).
3. Model spatial structure.
4. Define search strategy.
5. Predict values at unsampled locations.
6. Quantify uncertainty of the predictions.
7. Check that the model produces resonable results for predictions and uncertainties.
   No
   Yes
8. Use the information in risk analysis and decision making.

The first step, as in almost any data-driven study, is to closely examine the data. This typically starts by mapping the dataset, using a classification and color scheme that allow clear visualization of important characteristics that the dataset might present, for example, a strong increase in values from north to south or a mix of high and low values in no particular arrangement (possibly a sign that the data was taken at a scale that does not show spatial correlation).

The second stage is to build the geostatistical model. This process can entail several steps, depending on the objectives of the study (that is, the types of information the model is supposed to provide) and the features of the dataset that have been deemed important enough to incorporate. At this stage, information collected during a rigorous exploration of the dataset and prior knowledge of the phenomenon determine how complex the model is and how good the interpolated values and measures of uncertainty will be. In the figure above, building the model can involve pre-processing the data to remove spatial trends, which are modeled separately and added back in the final step of the interpolation process. It might also involve transforming the data so that it follows a Gaussian distribution more closely (required by some methods and model outputs). While a lot of information can be derived by examining the dataset, it is important to incorporate any knowledge you might have of the phenomenon. The modeler cannot rely solely on the dataset to show all the important features; those that do not appear can still be incorporated into the model by adjusting parameter values to reflect an expected outcome. It is important that the model be as realistic as possible in order for the interpolated values and associated uncertainties to be accurate representations of the real phenomenon.

In addition to pre-processing the data, it may be necessary to model the spatial structure (spatial correlation) in the dataset. Some methods, such as kriging, require this to be explicitly modeled using semivariogram or covariance functions, whereas other methods, such as Inverse Distance Weighting, rely on an assumed degree of spatial structure, which the modeler must provide based on prior knowledge of the phenomenon.

A final component of the model is the search strategy. This defines how many data points are used to generate a value for an un-sampled location. Their spatial configuration (location with respect to one another and to the un-sampled location) can also be defined. Both factors affect the interpolated value and its associated uncertainty. For many methods, a search ellipse is defined, along with the number of sectors the ellipse is split into and how many points are taken from each sector to make a prediction.

Once the model has been completely defined, it can be used in conjunction with the dataset to generate interpolated values for all un-sampled locations within an area of interest. The output is usually a map showing values of the variable being modeled. The effect of outliers can be investigated at this stage, as they will probably change the model's parameter values and thus the interpolated map. Depending on the interpolation method, the same model can also be used to generate measures of uncertainty for the interpolated values. Not all models have this capability, so it is important to define at the start if measures of uncertainty are needed. This determines which of the models are suitable.

As with all modeling endeavors, the model's output should be checked, that is, make sure that the interpolated values and associated measures of uncertainty are reasonable and match your expectations.

Once the model has been satisfactorily built, adjusted, and its output checked, the results can be used in risk analyses and decision making.

Variogram

A variogram is used to display the variability between data points as a function of distance. An example of an idealized variogram is shown below.

This variogram represents the variability between data points that lie along a 45 degree (+/- 10 degree) bearing from each other. Reading this variogram shows the following variability:

Point-to-point Distance	Variability
2.5 m	0.5
10 m	1.25
35 m	4.5
50 m	5.9

You might say that along this orientation, closely-spaced data points show a low degree of variability while distant points show a higher degree of variability. At some distance, in this case 72 meters, the differences between points will become fairly constant and the variogram will flatten out into a "sill." From 0 to 72 meters of distance, the data points can be considered "related," and this expanse is called the "range."

The "relatedness" of close points is not a huge surprise. But, are points more related in one direction than another? A similar variogram can be constructed for the same data set for data points that lie in a different direction from each other. For example, at a bearing of 135 degrees (+/- 10 degrees), the variogram might look like this:

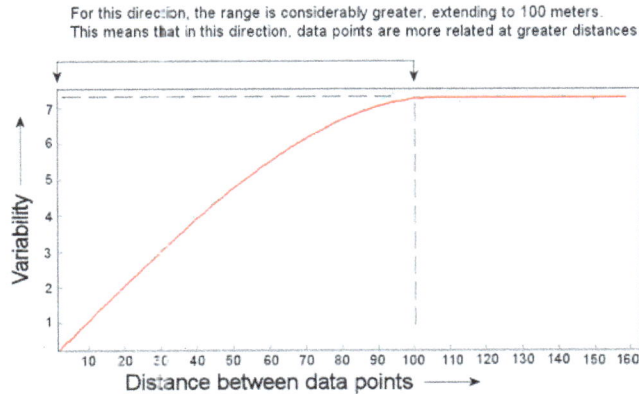

For this direction, the range is considerably greater, extending to 100 meters. This means that in this direction, data points are more related at greater distances.

Along this bearing:

Point-to-point Distance	Variability
2.5 m	0.25
10 m	1
35 m	3.5
50 m	4.5

The range for this variogram is 100 meters - considerably larger than the range in the previous example. This means that data points along this bearing can be considered to be more similar at greater distances from each other.

Significance of Variogram

Consider two synthetic data sets; we will call them A and B. Some common descriptive statistics for these two data sets are given in table:

Table: Some common descriptive statistics for the two example data sets.

	A	B
Count	15251	15251
Average	100.00	100.00
Standard Deviation	20.00	20.00
Median	100.35	100.92
10 Percentile	73.89	73.95
90 Percentile	125.61	124.72

The histograms for these two data sets are given in figures. According to this evidence the two data sets are almost identical.

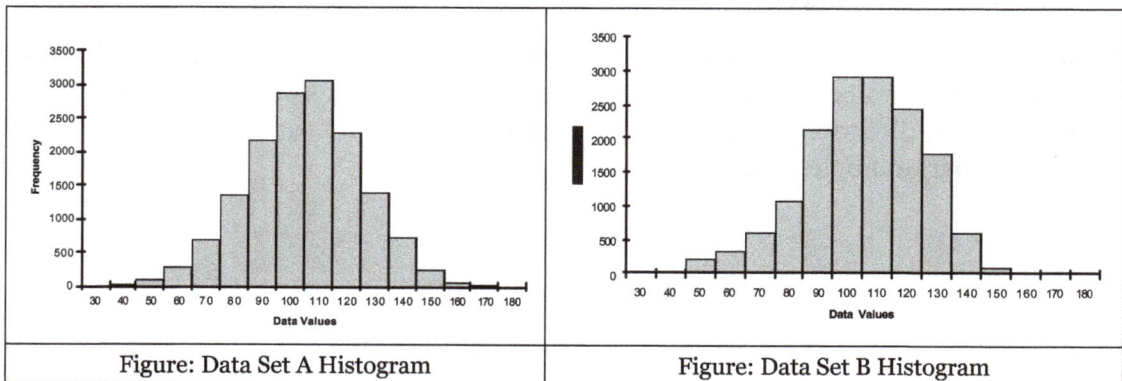

| Figure: Data Set A Histogram | Figure: Data Set B Histogram |

However, these two data sets are significantly different in ways that are not captured by the common descriptive statistics and histograms. As can be seen by comparing the associated contour plots, data set A is rougher than data set B. Note that we cannot say that data set A is "more variable" than data set B, since the standard deviations for the two data sets are the same, as are the magnitudes of highs and lows. The visually apparent difference between these two data sets is one of texture and not variability.

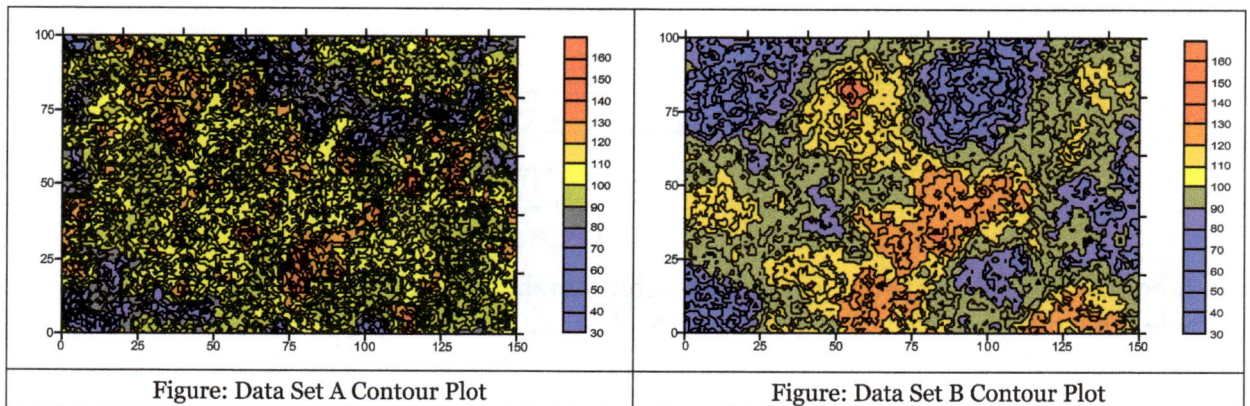

| Figure: Data Set A Contour Plot | Figure: Data Set B Contour Plot |

In particular, data set A changes more rapidly in space than does data set B. The continuous high zones (red patches) and continuous low zones (blue patches) are, on the average, smaller for data set A than for data set B. Such differences can have a significant impact on sample design, site characterization, and spatial prediction in general.

It is not surprising that the common descriptive statistics and the histograms fail to identify, let alone quantify, the textural difference between these two example data sets. Common descriptive statistics and histograms do not incorporate the spatial locations of data into their defining computations.

The variogram is a quantitative descriptive statistic that can be graphically represented in a manner which characterizes the spatial continuity (i.e. roughness) of a data set. The variograms for these two data sets are shown in figures. The difference in the initial slope of the curves is apparent.

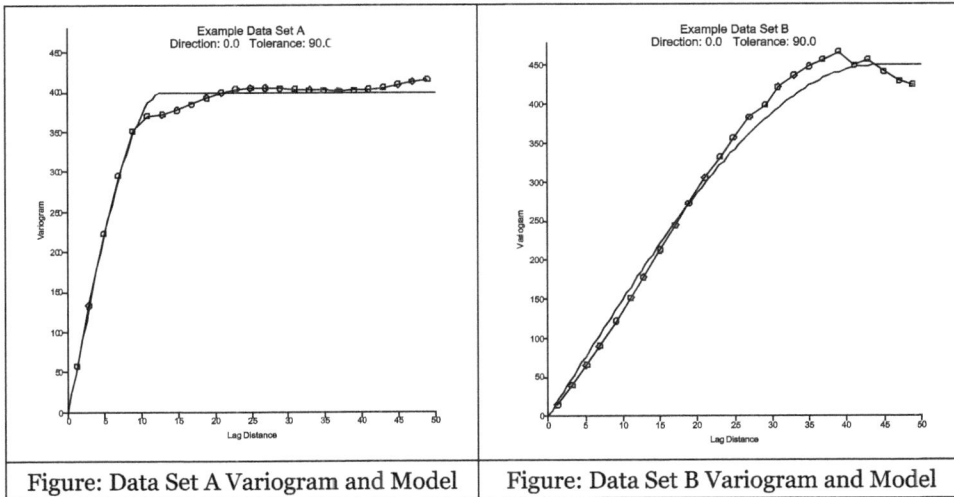

| Figure: Data Set A Variogram and Model | Figure: Data Set B Variogram and Model |

The mathematical definition of the variogram is:

$$\gamma(\Delta x, \Delta y) = \frac{1}{2}\varepsilon\left[\{Z(x+\Delta x, y+\Delta y - Z(x,y)\}^2\right]$$

Where, $Z(x,y)$ is the value of the variable of interest at location (x, y), and $\varepsilon[\ ]$ is the statistical expectation operator. Note that the variogram, $\gamma()$, is a function of the separation between points (Dx, Dy), and not a function of the specific location (x, y). This mathematical definition is a useful abstraction, but not easy to apply to observed values. Consider a set of n observed data: $\{(x_1, y_1, z_1), (x_2, y_2, z_2), \dots (x_n, y_n, z_n)\}$, where (x_i, y_i) is the location of observation i, and zi is the associated observed value. There are $n(n - 1)/2$ unique pairs of observations. For each of these pairs we can calculate the associated separation vector:

$$(\Delta x_{i,j}, \Delta y_{i,j}) = (x_i - x_j, y_i - y_j)$$

When we want to infer the variogram for a particular separation vector, $(\Delta x, \Delta y)$, we will use all of the data pairs whose separation vector is approximately equal to this separation of interest:

$$(\Delta x_{i,j}, \Delta y_{i,j}) \approx (\Delta x, \Delta y)$$

Let $S(\Delta x, \Delta y)$ be the set of all such pairs:

$$S(\Delta x, \Delta y) = \left\{(i,j)\big|(\Delta x_{i,j}, \Delta y_{i,j}) \approx (\Delta x, \Delta y)\right\}$$

Furthermore, let $N(\Delta x, \Delta y)$ equal the number of pairs in $S(\Delta x, \Delta y)$. To infer the variogram from observed data we will then use the formula for the experimental variogram.

$$\hat{\gamma}(\Delta x, \Delta y) = \frac{1}{2N(\Delta x, \Delta y)} \sum_{(i,j)\in S(\Delta x, \Delta y)} (z_i - z_j)^2$$

That is, the experimental variogram for a particular separation vector of interest is calculated by averaging one-half the difference squared of the z-values over all pairs of observations separated by approximately that vector.

Variogram Grid

If there are n observed data, there are n(n - 1)/2 unique pairs of observations. Thus, even a data set of moderate size generates a large number of pairs. For example, if n = 500, n(n - 1)/2 = 124,745 pairs. The manipulation of such a large number of pairs can be time consuming, even for a fast computer. Surfer pre-computes all of the pairs and stores the necessary sums and differences in the variogram grid. (Note: a variogram grid is not the same format as a grid used in creating a map.).

Variogram Models

We first generate a series of observed variograms for the raw data, calculating the variance between points at the specified distance increments and along each specified bearing.

The observed variograms, which represent your source data, are then fit to each of the 8 types of variogram models within the program. This is so that the data, which were sampled at discreet units, can be modeled as a continuous function, and the value for any unknown point at any distance can be interpolated. Once the best fit has been made, the "goodness of fit" is represented by a correlation coefficient for each variogram type. (Positive correlations indicate direct relationship between variance predicted by the variogram versus the lag variances with 1.0 being perfect. Negative correlations indicate inverse relationships. Zero indicates no relationship.)

- Spherical Models: The most commonly used model, with a somewhat linear behavior at small separation distances near the origin, but flattening out at larger distances and reaching a sill limit.

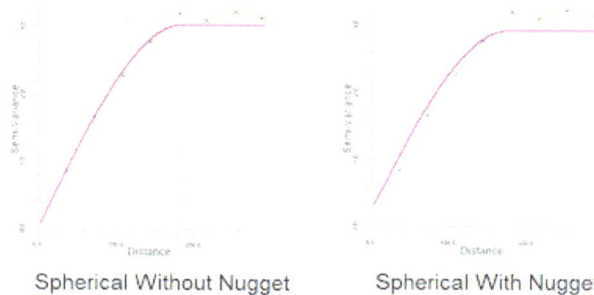

Spherical Without Nugget   Spherical With Nugget

- Exponential Models: Reach the sill asymptotically, with the practical range defined as that distance at which the variogram value is 95% of the sill. Like the spherical model, the exponential model is linear at small distances near the origin, yet rises more steeply and flattens out more gradually. Erratic data sets can sometimes be fit better with exponential models.

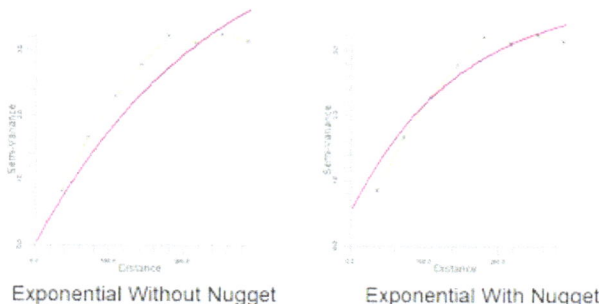

Exponential Without Nugget   Exponential With Nugget

- Gaussian Models: Characterized by parabolic behavior at the origin, then rising to reach its sill asymptotically at a practical range of 95% of the sill. This model is used to model extremely continuous phenomena. Care should be taken when using Gaussian model for kriging as they can produce somewhat erratic estimates, especially in areas of sparse control.

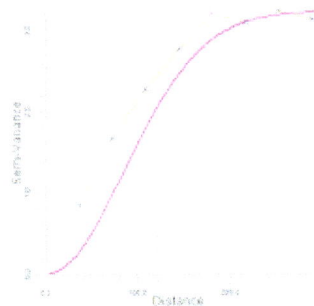

Gaussian Without Nugget   Gaussian With Nugget

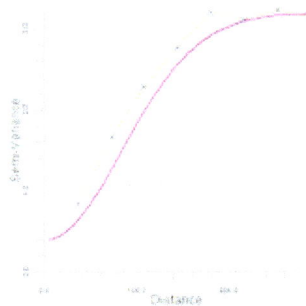

- Linear Models: Will show a sill within the data if there is a leveling off of the variance within the observed data, otherwise, the sill is assigned beyond the maximum distance.

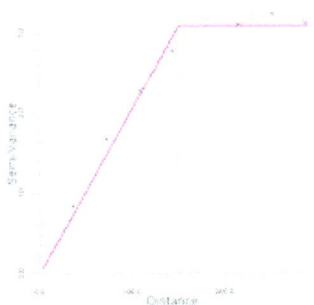

Linear Without Nugget   Linear With Nugget

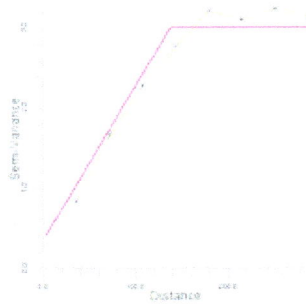

- "Nugget": In models with the "nugget" in effect, the variance does not go through the origin of the plot, indicating that even at very close distances (indeed, even at a distance of zero) the data points show some degree of variability. This can happen when, change occurs over the surface at distances less than the sampling interval. Variogram models with the nugget effect allowed can be used to illustrate the margin of error in your data.

One way you might conceptualize the nugget effect is to imagine that you are measuring geochemical data, and the size of your rock samples are about 1 mm in diameter. At very close distances, let's say 1 mm, the samples are likely to be very similar. On a variogram plot, the variance will probably go through the origin, and there will be no nugget effect. If, in contrast, your samples are 10 cm in diameter, the smallest sampling distance you can even get between points is 10 cm - a distance over which there may be much greater variability. The size of the sample or "nugget" itself gets in the way of determining variability at very close distances. By enlarging the sample size ot a watermelon or a Volkswagen, you may begin to get the picture of the effect of the size of the nugget or the sampling interval on the appearance of the variogram, and its relation to the margin of error in your data.

Directionality Ellipse

Once the observed variograms have been fit to each of these variogram models, the correlation can be reported, and the ranges for each direction determined. One method for illustrating the comparative ranges in all directions in the directionality ellipse below. Note the directionality that becomes apparent when viewing data in this way. The "major axis" lies at 125.5 degrees and the "minor axis" at 35.5 degrees.

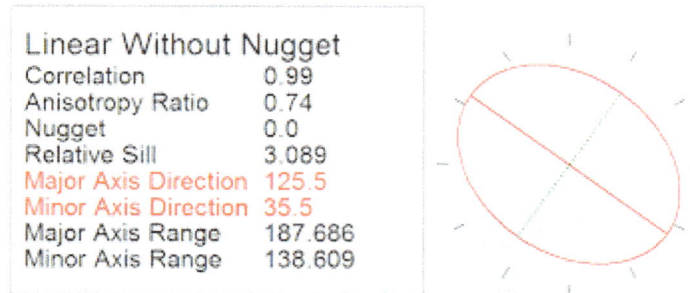

Linear Without Nugget	
Correlation	0.99
Anisotropy Ratio	0.74
Nugget	0.0
Relative Sill	3.089
Major Axis Direction	125.5
Minor Axis Direction	35.5
Major Axis Range	187.686
Minor Axis Range	138.609

Selecting the Variogram to use

The most obvious factor on which to make this decision is the correlation shown between a particular variogram and your data. The higher the correlation the better fit the more accurate the model and the more accurate the kriged grid. If you find that a number of variogram types show virtually the same goodness of it, then it's very likely that the grid that is kriged from any one of them will appear almost identical to the grid kriged from any of the others, making the decision rather arbitrary. Do note, though, that because the correlation coefficient is a representation of the correlation between the variogram and the lags, improperly selected lag dimensions (Manual kriging) won't necessarily be reflected by the correlation coefficient. In fact, very bad lag dimensions can produce high correlation coefficients.

One other factor you might consider is the effect of the nugget in the resulting grids and ultimate contour maps. The nugget effect illustrates the error in your date. When the nugget effect is allowed in a variogram, the resulting data grid may ultimately generate contours which do not honor the control points. It can't help it - it's simply showing you the error in your data. If you get such a contour map and if this is unacceptable, then you should opt for re-kriging the data grid using a variogram with a suppressed nugget.

Semi-variogram

Tobler's First Law of Geography states that "everything is related to everything else, but near things are more related than distant things."

In the case of a semi-variogram, closer things are more predictable and has less variability. While distant things are less predictable and are less related.

For example, the terrain one meter ahead of you is more likely to be similar than 100 meters away.

Semi-variogram charts out this critically important concept of how sample values (pollution, elevation, noise, etc.) vary with distance.

Soil Moisture Samples

Our example contains 73 soil moisture samples in a 10 acre field. In the north-west corner, the samples are much wetter with higher water content. But in the eastern quadrant, they are much dryer as color-coded in the image below:

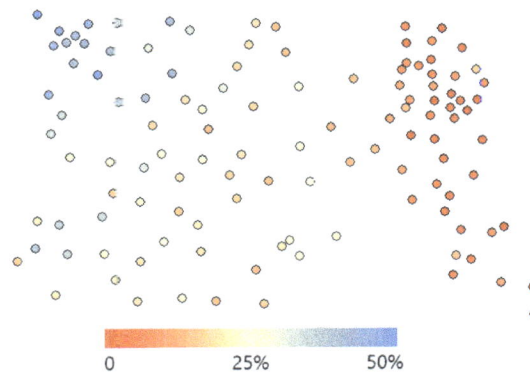

- How predictable are values from place to place?
- Are known values closer together more similar than values farther apart?

This idea can be described with statistical dependence or autocorrelation. Further to this, auto-correlation (things closer together are more similar than things farther apart) provides valuable information for prediction.

Working of Semi-variograms

To understand spatial dependence, you can estimate it with a semi-variogram. Semi-variograms take 2 sample locations and calls the distance between both points h.

In the x-axis, it plots distance (h) in lags, which are just grouped distances. Taking each set of 2 sample locations, it measures the variance between the response variable (water content in soil) and plots it out in the y-axis.

Depending on the observer, semi-variograms look like a big mess of points. For example, our soil moisture plot looks like this:

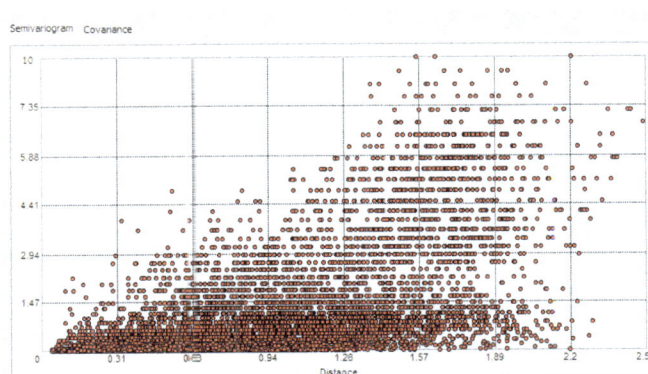

But you can really do some detective work by selecting individual points. When you take this single point on the semi-variogram:

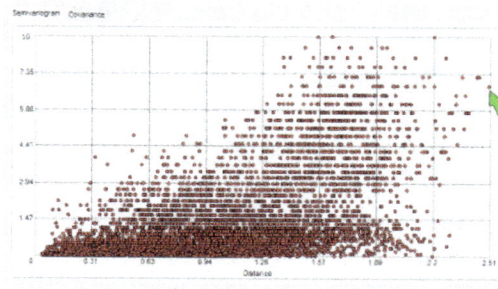

You can see which 2 points they represent on the map. This makes sense because they are a far distance apart from each other. Hence, its far-right position in the semi-variogram. It's actually this point highlighted below:

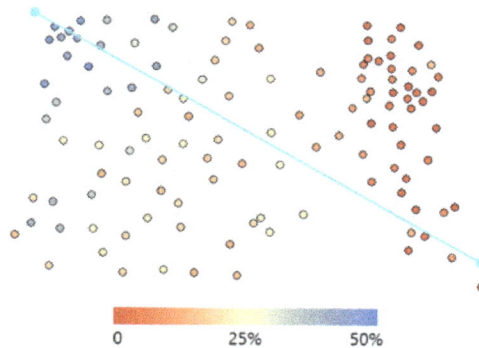

They also have a large difference from the mean value in that particular lag distance. It's positioned higher on the y-axis if the semi variance is high. As you probably noticed, the semi variance is smaller at closer distances and increases with larger lag distances.

We are looking at all distance between 2 samples and their variability. A semi-variogram considers all points and their distance with variance. That's why semi-variograms have so many points on it. Here's a subset of the data set above to see all the different sets of points that are being plotted out in a semi-variogram.

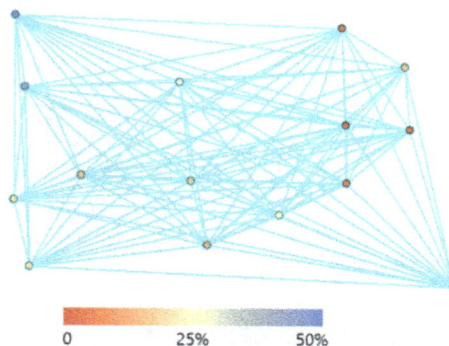

Different Sets of Points that are being Plotted out in a Semi-variogram.

Range, Sill and Nugget in Semi-variograms

At sample points with close distances, the difference in values between points tend to be small. In other words, the semi-variance is small.

But when sample point distances are farther away, they are less likely to be similar. This means that the semi-variance becomes large.

As distance increases away from sample points, there is no longer a relationship between the sample points. Their variance begins to flatten out, and sample values are not related to one another.

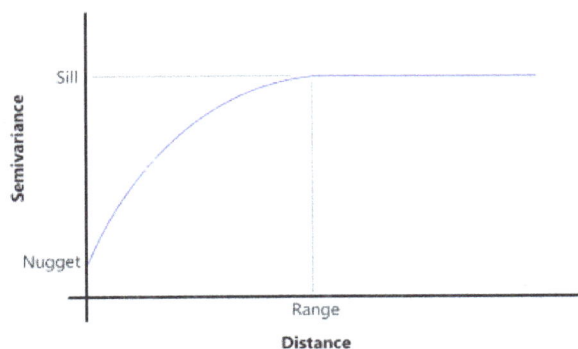

Sill: The value at which the model first flattens out.

Range: The distance at which the model first flattens out.

Nugget: The value at which the semi-variogram (almost) intercepts the y-value.

When you have two sample points at the same location, it is expected to have the same value so the nugget should be zero. Sometimes they don't and this adds randomness. But before the graph starts leveling, these value are spatially auto correlated.

As expected, when distance increases, the semi variance increases. There are less pairs of points separated by far distances, hence the less correlation between sample points.

But as indicated in the semi-variogram with the sill and range, it begins to reach its flat, asymptotic level. This is when you try to fit a function to model this behavior.

Mathematical Function and Models

You select the type of model for how it fits the data because it will provide a mathematical function to the relationship between values and distances. We use functions that are the best fit like exponential, linear, spherical and Gaussian.

Ideally, you are trying to lower your R-squared value, as best fit as possible. However, when you have an understanding of how the phenomena behaves with distance, you can better choose which model to use.

For example, here are the mathematical functions you can apply to semi-variograms.

Linear Models

A linear model means that spatial variability increases linearly with distance. It's the most simple type of model without a plateau, meaning that the user has to arbitrarily select the sill and range.

Spherical Models

The spherical model is one of the most common models we use in variogram modelling. It is a modified quadratic equation where spatial dependence flattens out as the sill and range.

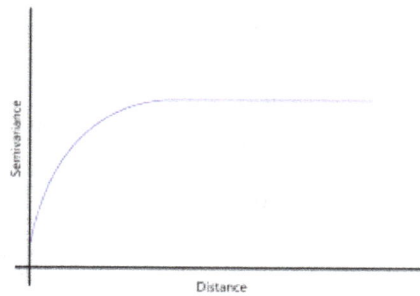

Exponential Models

The exponential model resembles the spherical model in that spatial variability reaches the sill gradually. The relationship between two sample points decay gradually, while at a distance of infinite spatial dependence dissipates.

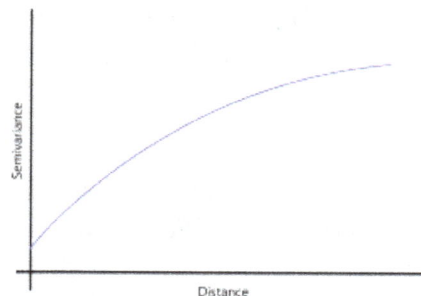

Gaussian Models

The Gaussian function uses a normal probability distribution curve. This type of model is useful where phenomena are similar at short distances because of its progressive rise up the y-axis.

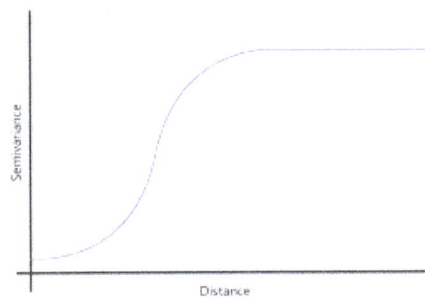

Circular Models

This type of prediction model uses a circular function to fit spatial variability in a semi-variogram. It resembles the speherical model function where spatial dependence fades away at its asymptotic level.

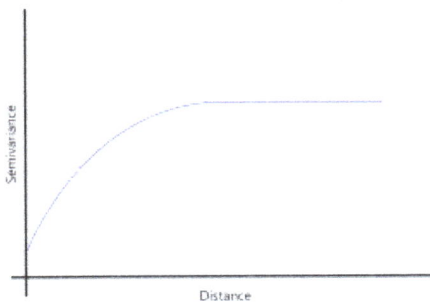

In conclusion, semi-variograms provide a useful preliminary step in understanding the nature of data.

Each phenomenon has its own semi-variogram and own mathematical function. The user uncovers the relationship between values and distances and then chooses the best fitting model.

Although semi-variograms are handy for understanding variation with distance, the model you choose from semi-variograms commonly goes into kriging. Because this type of interpolation technique uses the mathematical model from the semi-variogram, it's one of the best forms of prediction today.

This is because the variogram model influences the prediction of those unknown values during kriging interpolation.

Kriging

Kriging is an advanced geostatistical procedure that generates an estimated surface from a scattered set of points with z-values. Unlike other interpolation methods in the interpolation toolset, to use the kriging tool effectively involves an interactive investigation of the spatial behavior of the phenomenon represented by the z-values before you select the best estimation method for generating the output surface.

The IDW (inverse distance weighted) and Spline interpolation tools are referred to as deterministic interpolation methods because they are directly based on the surrounding measured values or on specified mathematical formulas that determine the smoothness of the resulting surface. A second family of interpolation methods consists of geostatistical methods, such as kriging, which are based on statistical models that include autocorrelation—that is, the statistical relationships among the measured points. Because of this, geostatistical techniques not only have the capability of producing a prediction surface but also provide some measure of the certainty or accuracy of the predictions.

Kriging assumes that the distance or direction between sample points reflects a spatial correlation that can be used to explain variation in the surface. The kriging tool fits a mathematical function to a specified number of points, or all points within a specified radius, to determine the output value for each location. Kriging is a multistep process; it includes exploratory statistical analysis of the data, variogram modelling, creating the surface, and (optionally) exploring a variance surface. Kriging is most appropriate when you know there is a spatially correlated distance or directional bias in the data. It is often used in soil science and geology.

Kriging Formula

Kriging is similar to IDW in that it weights the surrounding measured values to derive a prediction for an unmeasured location. The general formula for both interpolators is formed as a weighted sum of the data:

$$\hat{Z}(s_0) = \sum_{i=1}^{N} \lambda_i Z(s_i)$$

where:

$Z(s_i)$ = the measured value at the ith location;

λ_i = an unknown weight for the measured value at the ith location;

s_0 = the prediction location;

N = the number of measured values.

In IDW, the weight, λ_i, depends solely on the distance to the prediction location. However, with the kriging method, the weights are based not only on the distance between the measured points and the prediction location but also on the overall spatial arrangement of the measured points. To use the spatial arrangement in the weights, the spatial autocorrelation must be quantified. Thus, in ordinary kriging, the weight, λ_i, depends on a fitted model to the measured points, the distance to the prediction location, and the spatial relationships among the measured values around the prediction location.

Kriging Assumptions

The two main assumptions for kriging to provide best linear unbiased prediction are those of stationarity and isotropy, though there are various forms and methods of kriging that allow the strictest form of each of these assumptions to be relaxed:

- Stationarity – the joint probability distribution does not vary across the study space. Therefore, parameters (such as the overall mean of the values, and the range and sill of the variogram) do not vary across the study space. The same variogram model is assumed to be valid across the study space.

- Isotropy – uniformity in all directions.

Types of Kriging

There are several sub-types of kriging, including:

- Ordinary kriging: for which the assumption of stationarity (that the mean and variance of the values is constant across the spatial field) must be assumed. This is one of the simplest forms of kriging, but the stationarity assumption is not often met in applications relevant to environmental health, such as air pollution distributions.

- Universal kriging: which relaxes the assumption of stationarity by allowing the mean of the values to differ in a deterministic way in different locations (e.g. through some kind of spatial trend), while only the variance is held constant across the entire field. This second-order stationarity (sometimes called "weak stationarity") is often a pertinent assumption with environmental exposures.

- Block kriging: Which estimates averaged values over gridded "blocks" rather than single points. These blocks often have smaller prediction errors than are seen for individual points.

- Cokriging: in which additional observed variables (which are often correlated with each other and the variable of interest) are used to enhance the precision of the interpolation of the variable of interest at each location.

- Poisson kriging: for incidence counts and disease rates.

Sculpting a Prediction Model with Kriging

To really understand kriging, you have to know what interpolation is. As with all interpolation, we're predicting unknown values at other locations.

With an interpolation method like inverse distance weighting, you are making predictions without saying how certain you are.

Here's an example:

We predict the purple point, by taking an inverse weighted distance of the closest three input points (The values of 12, 10 and 10). Based on the distance, we calculate how far each input point and get a value of 11.1.

$$((12/350) + (10/750) + (10/850)) / ((1/350) + (1/750) + (1/850)) = 11.1$$

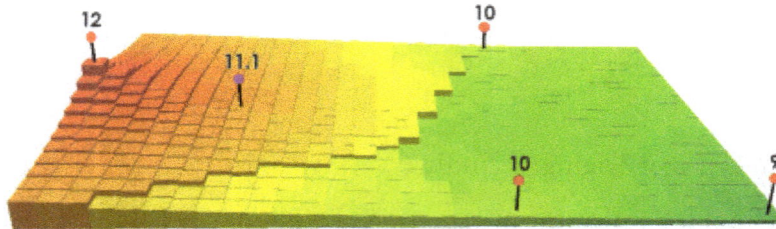

This is exactly how deterministic interpolation works. Simply, it uses a predefined function and it is what it is.

But it doesn't tell you how sure you are.

Kriging Interpolation

If a weatherman makes a forecast saying it's going to rain tomorrow, how sure are you that it's going to rain?

In other words, Instead of only saying here's how much rainfall at specific locations, kriging also tells you the probability of how much rainfall at a specific location.

You use your input data to build a mathematical function with a semi variogram, create a prediction surface and then validate your model with cross-validation.

Not only does geostatistics provides an optimal prediction surface, it delivers a measure of confidence of how likely that prediction will be true.

Meanwhile, kriging can generate the prediction surfaces and surfaces that describe how well your model predicts:

- Prediction: This surface straight predicts the values of your variable you are kriging.

- Error of Prediction: If depicts the standard error with higher standard of error where there isn't as much input data.

- Probability: The probability surface highlights when it exceeds a threshold.

- Quantile: This surface represents a best or worst case scenarios as a 99th percentile.

Semi Variogram as the Key to Kringing

Kriging relies on the semi-variogram. In simple terms, semi variograms quantify autocorrelation because it graphs out the variance of all pairs of data according to distance.

Chances are that closer things are more related and have small semi-variance. While far things are less related and have a high semi-variance.

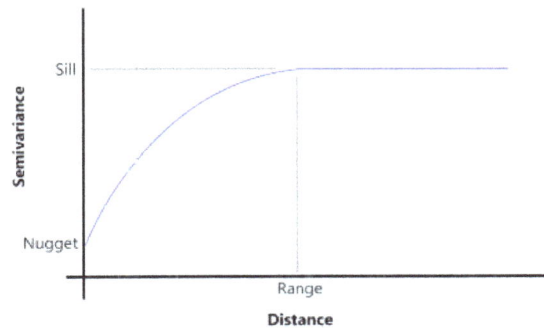

The purpose here is to fit a surface such as a polynomial that models the overall large-scale trend. Then, around that trend, we have variability with residuals where kriging comes in.

Based on your semi variogram results, you can select a semi variogram that is spherical, circular, exponential, Gaussian or linear. Alternatively, if you can make an intellectual justification for a mathematical model, then you pick that one.

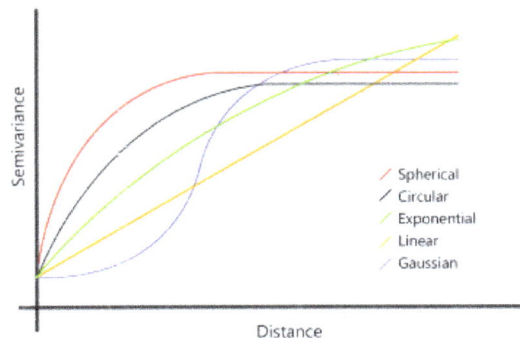

Before you even Begin, Check your Data

Before you even start kriging, your data needs to fit this criteria prior to ordinary kriging.

Kriging is the optimal interpolation technique if your data meets certain criteria. But if they don't meet that criteria, you can massage it or choose a different interpolation technique altogether.

- Your data needs to have a normal distribution

- The data needs to be stationary

- Your data cannot have any trends.

The following steps are ways to check your data to see if they fit this criteria. First, we suggest to plot out your points and symbolize, them from low to high. In our example, we use soil moisture samples taken in an agriculture field.

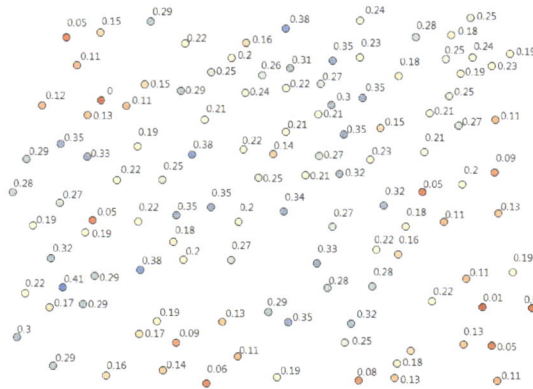

Assumption 1: Your Data has a Normal Distribution

While we are not exploring the spatial properties in this test, we are only checking that the values are fairly normally distributed. In other words, do the values of your data fit a bell-curve shape?

One of the ways to explore this is using a histogram.

At this point, you can check the histogram for any outliers and how much it looks like a bell-curve. In our case, it looks like it has a has a fairly good normal distribution.

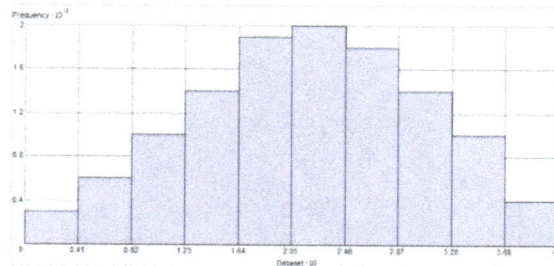

Alternatively, you can check your data with a normal QQ Plot. A normal QQ Plot compares how your data lines up with normally distributed data. If all points have a perfectly normal distribution, all your points would fall on the 45° line. In our case, the data follows a straight line.

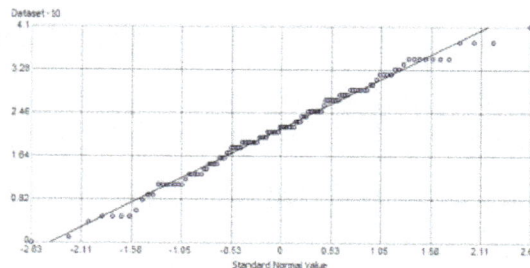

What if your data doesn't have a normal distribution? In this case, you will have to apply a transformation such as a log or arc sin until it becomes normal. Instead of selecting your own transformation, you can do a normal score transformation which pretty much does a lot of the work for you.

Assumption 2: Your Data is Stationary

What does it mean that your data has to be stationary? It means that local variation doesn't change in different areas of the map. For example, 2 data points 5 meters apart in different locations should have similar differences in your measured value. The variance is fairly constant in different areas of the map. Kriging is not optimal for abrupt changes and break lines.

You can check your data's stationarity with a voronoi map symbolizing by entropy (variation between neighbors) or standard deviation and look for randomness.

In our case, we do see some small amounts of clustering. Overall, for entropy and standard deviation voronoi maps show the data set is looking adequately stationary.

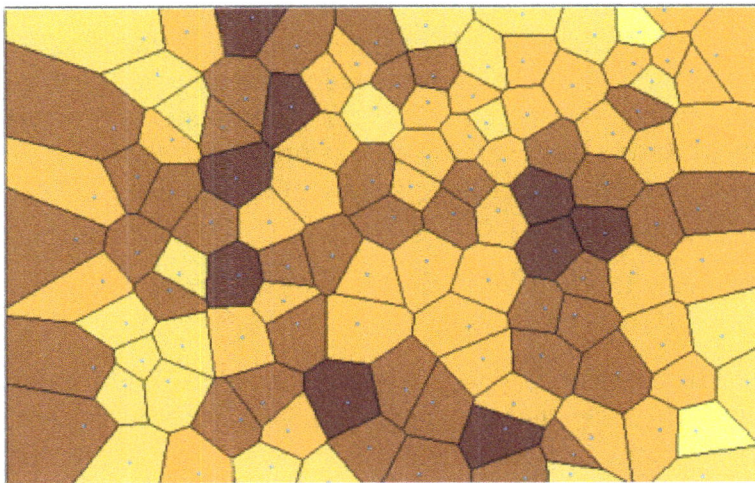

What do you do if your data isn't stationary? Empirical Bayesian Kriging (EBK) can help by treating local variance separately. Instead of variance being similar in a whole extent, EBK performs kriging as a separate underlying process in different areas. It still performs kriging, but it is done locally.

Assumption 3: Your Data doesn't have Trends

Trends are systematic change in data across an entire study area. We can check the trend analysis with the ESDA tool.

The green line shows the trend in the east-west direction, and the blue line depicts the trend in the north-south direction. Generally, we have higher soil moisture values in the center. But there's not enough of a trend in our data that it needs to be removed.

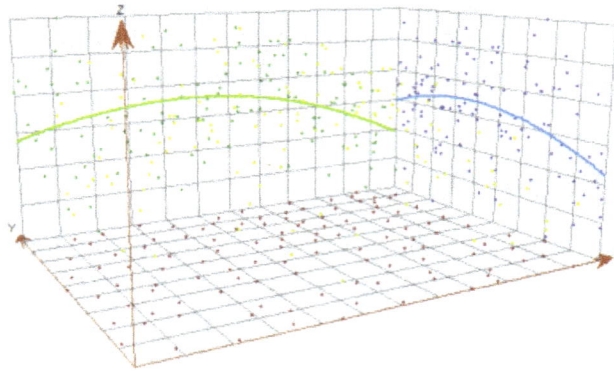

Markov Chain Geostatistics

Markov chain geostatistics is also called "the Bayesian Markov chain random field (MCRF) approach" or simply "the MCRF approach". It includes the MCRF theory, specific MCRF/coMCRF models, simulation algorithms, and transiogram modeling methods. This Bayesian geospatial statistical approach was initially proposed in Li and Zhang (2008) and Li (2007a), and then gradually developed and extended during last ten years. It was proposed mainly for modeling categorical/ discrete fields (e.g. various landscape classes) in multiple (two, three or spatiotemporal) dimensions. The core ideas of this approach include:

(1) Single-Markov-chain random field, which extends a conventional Markov chain into a locally-conditioned Markov chain (moving or jumping in a space);

(2) Spatial sequential Bayesian updating over nearest data within a neighborhood for local conditioning, which makes the locally-conditioned Markov chain essentially a special spatial Bayesian network;

(3) Spatial conditional independence assumption of nearest data within a neighborhood by extending the conditional independence property of Pickard random fields, which simplifies the MCRF general full solution (with multi-point likelihoods) into a simplified solution (with only transition probabilities);

(4) transiogram concepts and methods for transition probability model estimation from sample data and convenience of description, based on properties of transition probabilities and conventional stationary Markov chain theory.

Therefore, the MCRF theory formally extended the conventional 1-D Markov chain model into a multi-D spatial model, which has been proved to be practical in application studies. Although the name of MCRF may sound like a Markov random field (MRF) model, it is neither a conventional MRF model nor derived from the MRF model or Gibbs distribution. It is a spatial model with sequential Bayesian updating on nearest data within a neighborhood, and can be visualized as a probabilistic directed acyclic graph. Thus the MCRF model can be regarded as a special Bayesian Network over spatial data (or spatial Bayesian network).

Markov Chain Random Field

The MCRF theory proposed a special MRF where a single Markov chain moves or jumps to any location in a space, interacting with its nearest known neighbors in different directions, and its state at that location is decided by the interactions between the Markov chain and its nearest known neighbors. Thus, it theoretically extends a one-dimensional Markov chain into any dimensions for spatial modeling. The general solution in the conditional probability distribution of a MCRF Z at an unknown location u was derived as:

$$\Pr(Z(u)=k \mid Z(u_1)=1_1,...,Z(u_m)=1_m)$$

$$= \frac{\prod\limits_{i=2}^{n} P_{kl_i}^i(h_i).P_{l_1 K}^1(h_1)}{\sum\limits_{f=1}^{n}\left[\prod\limits_{i=2}^{m} P_{fl}^i(h_i).P_{l_1 f}^{1}(h_1)\right]}$$

(Li, 2007a), where $P_{kl_i}^i(h_i)$ represents a transition probability in the ith direction from state k to state l_i with a lag h_i; u1 represents the neighbor from or across which the Markov chain moves to the current location u; m represents the number of nearest known neighbors; k, l_i, and f represent states in the state space S = (1, ..., n); h_i is the distance from the current location to its nearest known neighbor u_i. With increasing lag h, any $p_{kl(h)}$ forms a transiogram, which represents the spatial (auto or cross) correlation of classes. It can be seen that the conditional probability distribution equation of a MCRF is actually composed of transiograms.

In practical use, however, the above general solution cannot be simply used directly, because it is necessary to consider the conditional independence of nearest known neighbors in cardinal directions for an optimal simulation. Normally in a rectangular lattice, four orthogonal cardinal directions are considered. So if only nearest data locations in four cardinal directions are considered, the MCRF model in above equation is simplified as:

$$\Pr(Z(u)=k \mid Z(u_1)=1, Z(u_2)=m, Z(u_3)=q, Z(u_4)=0)$$

$$\frac{P_{KO}^4(h_4) \cdot P_{kq}^3(h_3) \cdot P_{km}^2(h_2) \cdot P_{1K}^1(h_1)}{\sum\limits_{f=1}^{n}\left[p_{fo}^4(h_4) \cdot p_{fq}^3(h_3) \cdot p_{fm}^2(h_2) \cdot p_{1f}^1(h_1)\right]}$$

Where, 1, 2, 3 and 4 represent the four cardinal directions considered. In directions 2, 3, and 4, transitions are from the current unknown location u to its nearest known neighbors, but in direction 1 (i.e., the coming direction of the Markov chain), the transition is from the nearest known neighbor u_1 to the current location u.

In a simulation process, because the nearest known neighbors found within a search radius may not always reach four, the needed MCRF models may be further simplified from above equation.

It should be noted that the MCRF theory not only extends a single Markov chain for multi-dimensional simulation, but also resolves the small class underestimation problem, which occurred in some multiple-chain-based Markov chain models and was regarded as a "long-standing ghost" in Markov chain modeling. Li (2007a) found that the exclusion of unwanted transitions was the cause of the problem and the MCRF theory provided the way to avoid unwanted transitions.

Transiogram

The transiogram refers to a transition probability diagram for characterizing inter-class or intra-class correlations. A transiogram can be represented as a transition (or conditional) probability function on a continuous lag h:

$$P_{ij}(h) = \Pr(Z(x+h) = j \mid Z(x) = i)$$

Where, Z is a random variable and x represents one specific location. The second-order stationarity assumption is applicable here so that $P_{ij}(h)$ is dependent only on h, not x. An auto-transiogram $P_{ii}(h)$ represents the self-dependence (i.e., auto-correlation) of a single class i and a cross-transiogram $P_{ij}(h)$ (i ≠ j) represents the cross-dependence of class j on class i. Here class i is called a head class and class j is called a tail class. Apparently, the transiogram is similar to the indicator variogram in representing spatial continuity of categorical variables.

Practically, an experimental transiogram is directly estimated from sample data by counting the transition frequency from a class to itself or another class with different lags (e.g., numbers of pixels for raster data). If anisotropies are considered, experimental transiograms have to be estimated directionally, similar to estimation of variograms.

Transiogram Modeling

There are two methods to acquire continuous transiogram models. The first one uses mathematical models to jointly fit experimental transiograms. The joint modeling requires experimental transiograms be fit subset by subset and for each subset one transiogram model be inferred by the following equation:

$$P_{ik}(h) = 1 - \sum_{\substack{j=1 \\ j \neq k}}^{n} P_{ij}(h)$$

where, n is the number of classes, i is the head class, and $P_{ik}(h)$ is the inferred model. The above equation can guarantee that all transiogram models headed by the same class meet the summing-to-one condition at any specific lag. To guarantee the non-negative constraint and well-fitting of the transiogram model calculated by above equation other fitted transiogram models may need to be adjusted repeatedly.

Some basic mathematical models for modeling experimental transiograms were provided in Li. Figure shows that an experimental auto-transiogram and an experimental cross-transiogram are approximately fitted by basic mathematical models. This method is relatively time-consuming when the number of classes is large, but it permits incorporation of expert knowledge in estimation of transiogram models. Here expert knowledge refers to the knowledge of experts in parameter estimation of transiogram models, which typically include sills, ranges and model types (e.g., exponential, spherical). Therefore, this method is more flexible and widely applicable, particularly when samples are sparse and cannot provide reliable experimental transiograms.

The second one interpolates experimental transiograms into continuous models. This method is efficient but eliminates the chance of incorporating expert knowledge. Therefore, this method is suitable only when samples are sufficient and experimental transiograms are reliable. The linear interpolation method suggested by Li and Zhang (2005) is given as the following equation:

$$X = \frac{A.(D_B - D_X) + B.(D_X - D_A)}{D_B - D_A}$$

Where, A and B are the values of two neighboring points in an experimental transiogram D_A and D_B are the corresponding lags of the two neighboring points with $D_g > D_A$, and X is the value to be interpolated at a lag D_X between D_A and D_B $A_X A_B$. This linear method does not smooth experimental transiograms. It may be better to find some suitable methods that can also smooth experimental transiograms.

Simulation Algorithms

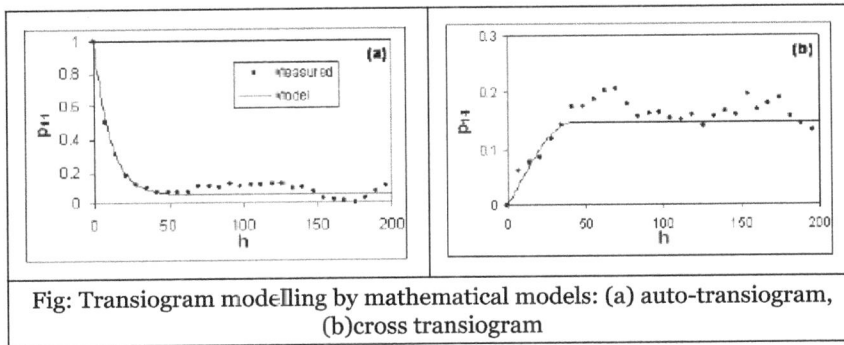

Fig: Transiogram modelling by mathematical models: (a) auto-transiogram, (b)cross transiogram

Markov chains are typical unilateral processes. It is widely known that unilateral process models (e.g., autoregressive processes, Markov chains, and Markov mesh models) usually generate inclined patterns (i.e., diagonal trends) due to their asymmetric neighborhoods used in multi-dimensional simulations. The way to avoid this problem is to design symmetric or quasi-symmetric paths. We have developed both fixed and random paths for multi-dimensional Markov chain simulation, which are symmetric or quasi-symmetric and can generate realistic patterns.

Fixed Path

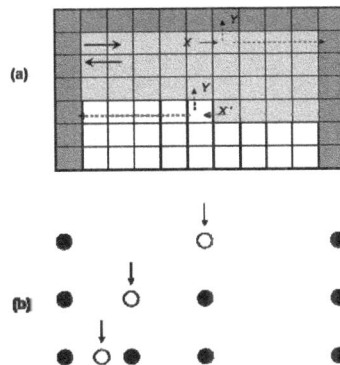

Figure: Fixed paths: (a) the alternate advancing path, (b) the middle insertion path

We first developed fixed paths for Markov chain simulation. One is the alternate advancing (AA) path figure(a) below. It described in Li et al. (2004). This path is identical to the path used in the herringbone method of Sharp and Aroian (1985) for dealing with the pattern inclination problem in multi-dimensional autoregressive process simulation and the path used in the scanning scheme algorithm of Wu et al. (2004) for avoiding the same problem in Markov mesh simulation. This path requires the simulation must be performed row by row alternately along opposite directions, and therefore, is quasi-symmetric in the lateral direction. To account for the directional asymmetry of spatial patterns under study, two different Markov chains (i.e., transition probability matrices) in opposite lateral directions were used in simulation.

The second path we suggested is the middle insertion (MI) path figure below, as mentioned in Li and Zhang (2005). This path also needs simulation to be performed row by row. But in each row the current simulated pixel must be the middle one between two nearest known neighbours in the row. This path is completely symmetric in the lateral direction. Note: paths that are not symmetry or quasi-symmetry in the lateral direction (for row-by-row simulation) cannot solve the pattern inclination problem. These fixed paths are convenient for dealing with line sample data or regular point sample data. To deal with randomly located samples, extensive software development is necessary. Although random path is more useful for dealing with random samples, fixed paths are still useful, especially for subsurface characterization.

Random Path

A random-path Markov chain sequential simulation (MCSS) algorithm was developed and presented in Li and Zhang (2007). The MCSS algorithm differs from kriging simulation algorithms because it considers at most four nearest known neighbors, one at most from each of the four search sectors that equally divide a search circle figure below. The reason for using four sectors is because in a rectangular lattice nearest known neighbors in the four cardinal directions can be regarded conditionally independent. To cover the whole search circle a cardinal direction has to be replaced by a search sector. A random path is generally quasi-symmetric.

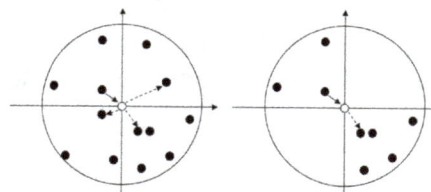

Figure: Random path

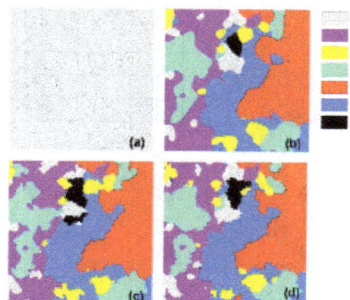

Figure: Simulted results by MCG using the random path, conditioned on a random land cover sample set of 499 points: (a) samples, (b) optimal map, (c) and (d) realizations

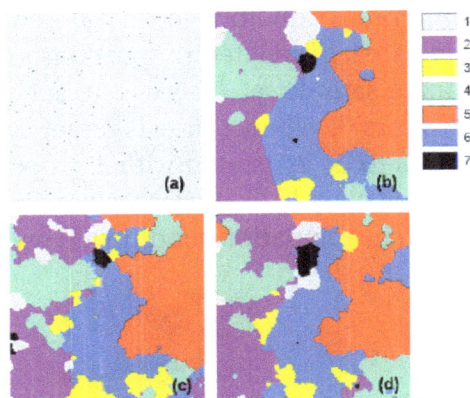

Figure: Simulated results by MCG using the random path, conditioned Conditioned on a random sample set of 130 Points: (a) samples, (b) optimal map, (c) and (d) realizations

Simulations and Analyses

Figure shows simulated results of seven land-cover classes using the MCSS random path algorithm. The simulation was conditioned on a random sample set of 499 points in a 35 km² area (a 295 by 295 lattice). Apparently the simulated pattern is polygonal. Different realizations are imitative of each other. The optimal map based on maximum occurrence probabilities is also imitative of simulated realizations. When the sample points are reduced to 130, simulated pattern becomes relatively simpler in the optimal map, but from the simulated realizations it can be seen that all classes are still fairly generated and realizations are still quite imitative of each other figure Compared to figure below, the pattern change in figure is due to the sparser sample set. This indicates that different sample sets provide different conditioning information and samples play the major role in determining simulated patterns in MCG.

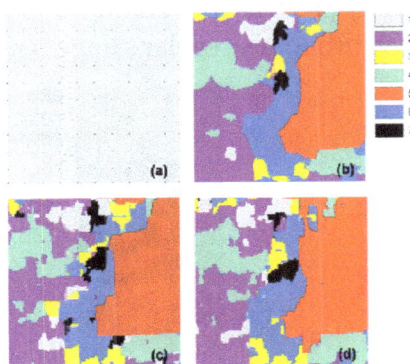

Figure: Simulated results by MCG using the AA path, conditioned on a regular sample set of 121 points: (a) samples, (b) optimal map, (c) and (d) realizations

Using a regular sample set of 121 points and a fixed path - the AA path, we also can get good results except that simulated polygons in realizations usually have uneven boundaries figure below Using such a sparse sample set as that in figure, small classes such as classes 1, 3 and 7 do not have the problem of being underestimated, which strongly occurred in some multiple-chain-based Markov chain models, as demonstrated in Li et al. (2004) and Zhang and Li (2007). Because small classes may account for different proportions in different sample sets, it is reasonable for them to be estimated differently in simulations conditioned on different sample sets, even those sample sets come from the same study area. Simulated proportions of classes are largely decided by their proportions in the conditioning sample set. There is no such issue called "underestimation of under sampled classes (or indicators)". When a class is under sampled relative to the expected proportion, it is correct for the class to be estimated with a proportion lower than the expected, as shown in Li (2007c), where a small class had a lower proportion in the sparser sample set and consequently had a lower estimation in simulation conditioned on the sparser sample set.

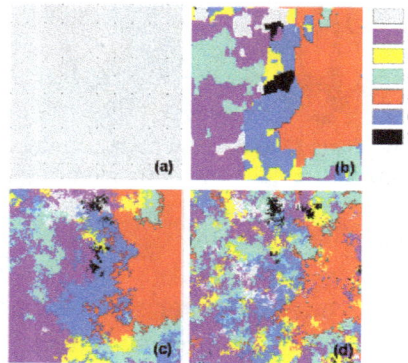

Figure: Simulated realizations by MCG and SIS, conditioned on a regular sample set of 121 points: (a) samples, (b) a realization from MCG using the AA path, (c) a realization from SIS-OIK, and (d) a realization from SIS-SIK

Because MCG uses nonlinear estimators and incorporates interclass relationships, it has advantages over indicator kriging in dealing with multinomial classes. From figure above, it can be seen that given the same conditioning sample set of land-cover classes, SIS-OIK (SIS with ordinary indicator kriging) and SIS-SIK (SIS with simple indicator kriging) generate dispersed patterns. The realizations generated by SIS-SIK apparently show a mess – different classes are mixed together. This means that without ancillary data SIS may not generate high-quality patterns in realizations to represent the spatial distribution of classes. Figure shows similar results with another regular sample set in soil types.

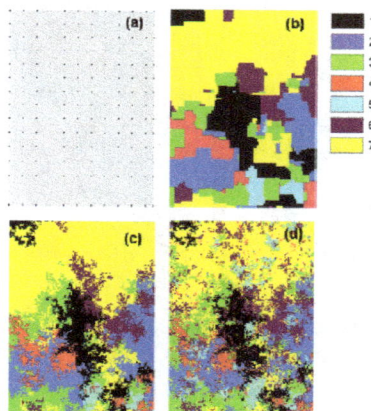

Fig: Simulated realizations by MCG and SIS, conditioned on a regular soil type sample set of 180 points: (a) samples, (b) a realization from MCG using the AA path, (c) a realization from SIS-OIK, and (d) a realization from SIS-SIK.

Polygonal patterns are in accordance with the custom of area-class mapping and are also convenient for human understanding and data processing using GIS tools. For example, they can be readily transformed into vector data. More importantly, many categorical variables such as lithofacies tend to have polygonal patterns in the real world because different facies usually have abrupt boundaries. Even for soil types and land cover classes, abrupt boundaries are easy to see in the natural world. Deutsch (1998) suggested using a post-cleaning method to remove the short-scale features in realizations from SIS. While the cleaning level is not easy to control without prior knowledge about the real pattern, MCG makes such post-cleaning unnecessary.

In conclusion, MCG has some advantages over conventional geostatistics in dealing with categorical variables because of its simplicity, nonlinearity and generalization. Currently, technical development in MCG is limited. Its application is mainly focused on simulation of categorical variables, particularly multinomial classes, without consideration of secondary variables.

Geodemographic Segmentation

Geodemographic segmentation is the statistical classification of people (demographics) based on where they live (geography).

The analysis relies on two primary principles:

1. People who live next to each other are more likely to share common characteristics than two people chosen at random.

2. Locations can be categorized by the characteristics and demographics of their residents. If they have a similar demographic makeup, multiple locations can fall into the same category even if they are separated by great distances.

Geodemography is often traced back to the work of famed London social reformer Charles Booth in the late nineteenth century. Skeptical of the existing data on poverty, Booth and a team of researchers set out to explore the role, impact, and distribution of poverty in London. The team effectively overlaid census data on top of a map of London to identify where, poverty clustered. The result of this research was the first "poverty map," a topographic visualization of poverty.

The analysis revealed that wealth clusters in specific areas—that is, immediate neighbors are likely to have similar income levels. The new data further revealed that the previous measures of poverty were severely understated and brought about a surge of social reforms, including old age pensions, free school meals, and other poor-relief policies.

While it started as a way to look at income distribution, location has since become a lens through which we can study any element of consumer behavior or demography. Just as income tends to cluster geographically, so do our lifestyles, interests and preferences.

Neighborhoods can be classified as predominantly homeowners or renters, blue collar or white collar, country music lovers or rock fans, young or old, white or black. It'll rarely be a perfect classification, of course. A predominantly Democrat neighborhood is bound to have a few Republicans; however, the differences within the neighborhood are fewer than those between other neighborhoods.

Thanks to the rise of location sharing and mobile technology, we're now even able to move beyond where someone lives to analyze their spatial behaviors—that is, how we move and interact with the world. We can now classify people by their shopping patterns ("people who shop at Whole Foods tend to live here and work here"), interests ("people who go to museums tend to live in higher-income neighborhoods"), and lifestyles ("people who to go a gym are more likely to rent than buy").

However you're looking to analyze consumers, location provides a valuable lens for identifying and grouping common characteristics.

Think of the World as a Website

If geodemographic segmentation is a difficult concept to grasp, consider something that's likely a little more familiar: Online marketing.

Customer segmentation is widely used in the world of online marketing. By slicing and dicing specific demographics and online behaviors, digital marketers have long been able to:

- Segment website visitors by behavior, looking at which pages they visited in order to send visitors more personalized communications and recommendations.

- Analyze conversion and site engagement metrics across cohorts grouped by age, gender, location, acquisition date, lifetime value, and website engagement.

- Dynamically personalize website content to send different messages to different types of visitors.

Geodemographic segmentation bridges the online and offline worlds, taking what digital marketers have been doing for years and putting that power in the hands of anyone looking at the physical world. Take a look at how these online and offline tactics compare:

With digital segmentation, you can...	With geodemographic segmentation, you can...
✓ Segment visitors by the types of content they read on your site.	✓ Segment people by the types of places they visit in the physical world.
✓ Classify visitors by their IP address.	✓ Classify individuals by their home address.
✓ Look at averages among segments or cohorts to see how your marketing resonates with different groups.	✓ Look at averages among neighborhoods to see which populations and demographics your marketing resonates with.
✓ Use user-generated data and browsing behavior to infer basic demographics, like age and gender, for an IP or user.	✓ Use census data to tap into a wealth of demographic attributes for a neighborhood or home address.
✓ Filter digital advertising to age, gender, and location.	✓ Filter local advertising to advanced demographics, like income, homeownership, race/ethnicity, and family structure, through the lens of location.
✓ Prioritize the segments (ideal customer persona) with the highest predicted lifetime value.	✓ Prioritize the segments (ideal customer persona) with the highest predicted lifetime value.
✓ Personalize content to different segments.	✓ Personalize content to different segments.

Putting Geodemographic Segmentation into Action

Today, the role of geodemographic segmentation has gone far beyond research and public policy. Marketers of every industry are realizing just how powerful of a lens location can be when it comes to understanding and engaging their customers.

Here are just a few strategies.

Plumbers and other Service Area Businesses can Target Key Neighborhoods

For a plumber, HVAC repair service, or other service area business (SAB), it can be a challenge to predict who will need your services. With little behavioral data to make sense of, SABs typically cast a wide net with their local advertising, targeting everyone within a certain number of miles. Of course, this also means they risk paying for a lot of clicks that won't convert into paying customers, such as apartment renters looking for quick tips while they wait on the rental company or low-income neighborhoods that might not be able to afford a higher-grade service.

This broad approach can also get expensive fast, especially with plumbing keywords costing north of $40 a click.

Geodemographic segmentation offers SABs a way to eliminate this ad waste by targeting locations characterized by certain census data:

- Homeownership, which can be used to exclusively target areas known for a high owner/renter ratio.

- Household value or income, which can be used to focus costly clicks on households likely able and willing to pay more for a high-quality service.

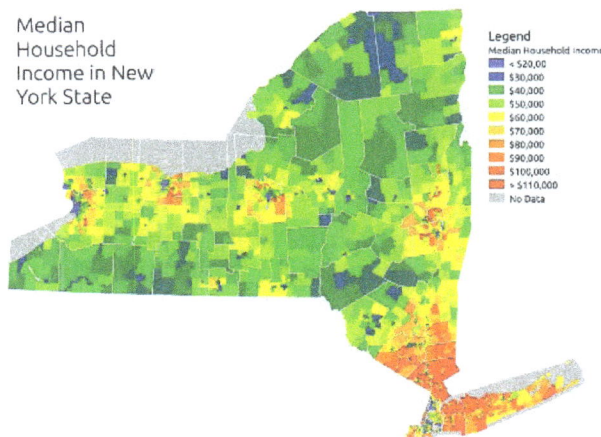

Segmenting New York by average household income segmenting by household income can be a great way to filter by socioeconomic attributes.

Political Campaign Managers can Target Down Party Lines

Political affiliation has often been the holy grail of political advertising. If campaign managers know where their party's voters live, they can focus their ad spend on those most likely to vote in their favour as opposed to having ads fall on deaf ears with the opposing party.

If you're able to get your hands on voter records, you can use geodemography to identify which locations are most popular with your party and restrict your ad spend to these areas. Alternatively, you can target swing counties by looking for areas with a roughly even distribution of political affiliations.

Some campaign managers are even using spatial data to target not just where people live, but where they visit. Using this strategy, advertisers can target key voter blocs, such as churchgoers (those recently observed at a church) or blue-collar workers.

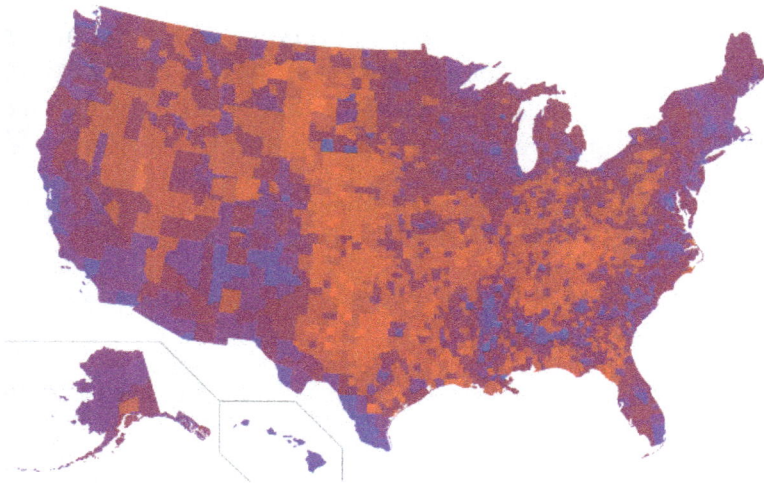

Segmenting by political affiliation can help politicians focus their campaigning on key swing counties and neighborhoods, highlighted here in purple.

Even e-commerce Companies can Leverage Geodemographic Segmentation

While geo-marketing is usually thought of in terms of brick-and-mortar or offline businesses, e-commerce websites are starting to invest more and more in offline data.

Instead of targeting nationwide to get to every possible customer, e-commerce companies are relying on geodemographic segmentation to highlight their most probable customers. Tapping into census data like income, racial composition, and family structure helps online brands pinpoint those locations that have residents who share the most in common with their ideal customer persona. This, in turn, allows them to spend much less while getting just as much, or even more, in return.

If, for example, you run an online store dedicated to new parents and know that your target market is in the mid-to-upper end of the income range, you can use geodemographic segmentation to target those locations with a high concentration of upper-income young parents. Both of these attributes are hard to pinpoint using online behaviors alone but can be readily found in location data (in this case, the census).

Principles Governing Geodemographic Segmentation

Two principles, on which geodemographic segmentation is based on, are:

- The first principle on which geodemographic segmentation is based is that people putting up in the same neighbourhood are inclined to have similar tastes and interests.

- The second principle on which geodemographic segmentation is based is that two neighbourhoods with similar characteristic features can be put into the same category. This is, irrespective of the distance between the two.

Advantages of Geodemographic Segmentation

There are several advantages of geodemographic segmentation:

- Effectiveness: This means that top companies can ensure that their marketing strategy is on point. Once the target group is defined, it is easier to know how to make the product fit into the outline of their needs. Thus, the effectiveness of the company is assured. The primary motive is to make the product sell by making the potential buyers believe that their needs can be fulfilled by the particular item. Consumer profiles are thus easier to understand and interpret.

- Ease of operations: This means that once the target group is known, accordingly steps can be taken to ensure that their attention is caught for the product. For example. In case of an item for kids, the advertisements can be vibrant and free toys may also be provided. On the other hand, if the target is the aged, then accordingly means may be adopted. It is very important to understand the taste of the potential clients. In fact, in order to ensure ease of operations, the marketing teams may also be divided accordingly and they may further appoint and allocate responsibilities. Thus, a hassle-free execution of the assigned task is made possible.

- Optimum utilisation of resources: Geodemographic segmentation ensures that there is no waste of time or resources. Once the target is known, steps of marketing are taken accordingly. There is no waste on the trial and error method where there is no surety of what is going to work and what isn't. Proper planning is done beforehand to save time, energy and efforts. There is a harmonious execution of all plans.

- Lack of pressure: This pertains to those who are working for the company. Proper and timely allocation of work helps to cut down on the workload. This means that everyone knows what their role is and what is expected of them. They can this focus on their own work and not be confused as to how things are to be done.

Disadvantages of Geodemographic Segmentation

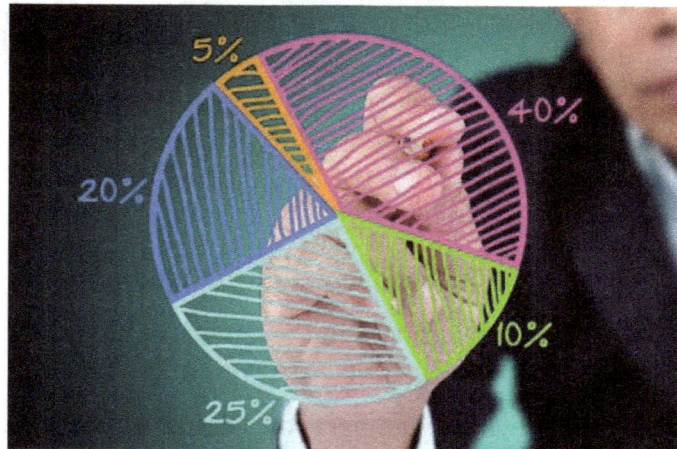

- The very assumption that all those in a particular group, are bound to have similar needs is a big disadvantage. It is not necessary that this will be the case. No two humans think alike or have the exact same needs. Thus, geodemographic segmentation leaves very little scope to cater to individual differences. In fact, they are totally ignored. This, it is not necessary that the segregation is bound to yield good results for the company. Everyone has a different lifestyle, daily situations to face and problems to solve. Thus, these are overlooked in case a company engages in geodemographic segmentation.

- Various tastes cannot be met through geodemographic segmentation. This is so because what may appeal to one person may be terrible in the eyes of someone else. At times, even if a product fits the needs of everyone, it fails to meet the tastes of people. This is where a matter of choice steps in. While the first disadvantage pertains to a compulsion, this is one where the masses have the freedom to pick. Thus, there may be times when competing brands and companies come up with something more appealing to the senses. A big factor regarding this could the pricing of a product. At the end of the day, the pocket pinch matters the most to the common man.

- At times, the information collected may be skewed. This means that it may be faulty or the population in one may consist of too many people of the same age or profession or other likewise element. For example, in a locality of factories, the main population will be labourers and the working class while students will be the main population in a region dominated by universities. Thus, the skewed population can yield skewed data. In such cases, there is an overall misinterpretation of the area. Other factors that lead to such errors may also be a migration of population and sex ratio between the males and females. Problematic geodemographic segmentation can, in turn, lead to reduced accuracy of operations and adoption of incorrect and ineffective marketing methods. This defeats the entire purpose of the task and proves to be a total waste.

Thus, geodemographic segmentation is significant and needs to be done strategically and wisely. Though it has both pros and cons, it can prove to be of great help to maximise the sales and profits incurred by the company.

Geotargeting

Geo-targeting or geotargeting refers to the practice of delivering different content or advertisements to a website user based on his or her geographic location. Geo-targeting can be used to target local customers through paid search campaigns.

Working Geotargeting

As anyone who works in marketing knows, location is one of the most important elements you can use to target an ad campaign. In traditional media, most geotargeting is implicit.

In digital however, you can't assume: YouTube and Facebook are just as easily accessed from Japan as they are from New York. But beyond controlling cost and ensuring ads only run within the footprint of where the advertiser actually sells their product, geotargeting offers huge opportunities for marketers.

For one, geotargeting in digital allows for more sophisticated measurement and personalization than was ever possible with traditional media. And with the proliferation of mobile devices, and the remarkable granularity and specificity the provide in terms of location, geotargeting has never been more powerful.

But while many marketers understand the value of geotargeting, not many are likely to understand how the technology behind it works. Having a firm grasp on the technology though is actually critical in this case, as different solutions take different approaches to the problem of determining the physical location of a consumer, and the simplest solution is usually the least accurate. Just as traditional strategies don't always translate well to digital, many desktop strategies don't translate well to mobile.

Geotargeting users Online

In online environments, ad servers look at a user's IP address to figure out their location. Behind the scenes, the ad server maintains a large database that has every IP address already mapped to its country, state, and postal code. So, when a request comes in, the ad server strips the IP address from the header of the request, queries this table, finds the necessary location data, and then picks an ad that matches that criteria.

Now the ad servers don't create this table themselves, they license it from another company like Max Mind or Digital Envoy, whose primary business is geo-location data. This is no enviable task; IP addresses themselves don't necessarily have a obvious pattern in the way they are assigned like a telephone area code would. It's a bit like solving a mystery, and the geo-location companies use a variety of methods to approach the problem.

The first thing to do is figure out which ISPs own what IP ranges. This is public information, used so that an IP isn't shared among many users across different ISPs. And since ISPs tend to serve a particular region, usually a country can be assigned to the user with this information alone.

Now, to figure out state and postal code involves a more complex process. At the core though, the geo-location services build up a network of servers from which they can send out pings, or connection requests, and known physical locations of public entities like universities and government office IPs. Eventually, with enough data, the geo-location company has the capability to triangulate any IP on the web.

It works like this – if there is an IP address the company wants to locate, they ping it from a few of their servers, for which they already know the location. A ping is just a way to test if a computer can connect, and how long it takes to do so, but doesn't transmit any meaningful data. Then, by looking at the time it takes each server to connect, it can establish a shared point or origin, and thereby physically locate the user. It uses the public IP locations to validate their approach and check for anomalies in network latency which would lead to bad data.

The risk to this approach is that it isn't always terribly accurate beyond the city to zip code level. If, for example, you were to use Max Mind's demo service to locate your own IP, it will likely show you perhaps a mile away from your actual address, likely at the nearest network node, the point at which your computer connects to your ISP's network infrastructure.

For this reason, some companies have taken a more direct measurement approach to IP geo-location vs. trying to infer it through ping triangulation. It's far more straightforward, but requires a lot more manual effort. Basically, these companies send cars out to drive up and down every street in the country and log WiFi IP addresses as well as their physical location to populate the same table that more traditional geo-location companies build through technical means. Google and Skyhook both use this approach.

Practical Tips for using Geo-location to Reach Target Audience

Find a Venue where your Target Audience will have Specific Wants or Needs

Stadiums, airports, universities, and malls are examples of specific venues that can be targeted in order to reach specific interest groups. Stadiums provide a great opportunity to focus on specific short engagement events with an audience defined by that event. They often host fans from two specific cities or schools or fans of a specific music genre that is heavy in one demographic. A band like One Direction, for example, is likely to attract school-age female fans.

Use these consumer characteristics to time and target your marketing. For example, airports on weekdays are a great source of business travelers looking for high-end restaurants, while weekends and Spring Break bring more leisure visitors and families looking for more casual dining options. Likewise, dance clubs and bars can benefit by promoting 18 and over events targeted at universities whose student bodies are largely between the ages of 18-21. These are just a few examples of how venues define audiences that can be effectively targeted.

Exclude Locations where your Target Audience will not be

Not only can you define an area you wish to reach, you can carve out an area you wish to exclude.

Exclusion can be done by venue or one side of the street or any area that could have been specifically targeted.

For example, clubs and bars that might otherwise want to target university students may exclude that same area during breaks or the summer when most students are away.

Excluding locations may also be a more cost-effective way to avoid the higher ad rates of high demand target areas. It is shown in the figure below how lower-cost, broader area ad campaigns can accomplish the same targeting goals by opting out of all areas but your desired target location.

Define a Radius by Distance or Time around your Store or Area of Interest

Geo-fencing allows marketers to set a perimeter around a physical location in which ads can be delivered. For geo-fencing ads, they may include creative messages acknowledging the user's location or may include location-based features such as a store locator.

For example, a coffee shop can set a 1-mile perimeter around its store and reach any user within that radius. Or, it could set a 3-mile perimeter around a nearby office complex to reach users that may be looking for somewhere to grab coffee before going into work. You can also try geo-conquesting, which targets customers around a competitor's location.

Another way to define a perimeter is not by distance, but by time. A company named iGeolise developed a platform they call Travel Time, an API that allows mobile apps and sites to search by time rather than distance. This could be useful for a condo unit near downtown looking to attract workers with very long commutes, or a restaurant targeting hotel patrons within a 10-minute walking distance.

Adjust Your Bid on Ads to Prioritize Better Locations

One concern with specific targeting is the loss in volume of audience. Even if you have an other-worldly 10% click-through rate, that's just 10 click throughs if only 100 people see your ad.

In low performing locations, the business developed from those areas may be outweighed by the campaign cost. By raising your bid for more desirable target locations, you increase your exposure in that area, while lowering your bid in other areas keeps your reach broad at a justifiable cost. These adjustments are a way of optimizing ad performance.

An event planning company or marketer for a musician that is hosting a concert in Chicago may use bid adjustments to prioritize Chicago, but also reach, at a lower cost, Milwaukee, WI and Grand Rapids, MI, both of which are driving distance.

Use Location-specific Keywords for Paid Search Ads

Geotargeting doesn't always mean you have to capture where someone is physically located. Consumer intent is conveyed all the time by search queries, and location is a commonly included term. Consumers often narrow their own searches by adding in the name of a city or district.

For example, "Austin gyms" or "coffee shops near Dupont Circle" or "uptown restaurants" provide location intent that you can target. Include location terms such as area code, ZIP code, neighborhood, community name, nearby landmarks, popular venues, tourist destinations, well known street names, local jargon and other keywords that will help you get found when a consumer is searching for businesses around you.

Predict your Audience by Geography

Geography can also be used to predict desirable demographics and information about users in that area. Neighborhoods can often be delineated by residents' income bracket, age, ethnicity, education, and many other demographics or interests. Politicians often draw district boundaries into areas of common political constituencies that also predict demographics or common values.

Knowing your business' target audience and matching it up with where they live or work helps you find those who might be most interested in your product or service. For example, a ticket broker might want to advertise NCAA basketball tickets in the state of Kentucky and might think of using Kentucky basketball in its messaging. However, Louisville basketball would be preferable for any advertising within 50 miles of the city on the Kentucky side of the border and 70 miles into Indiana due to the strength of Louisville's fan base in those areas.

Discover Location Intent by Search History

Targeting ads using search history allows marketers to deliver location specific ads to consumers, even if the consumer's tracked location doesn't match the physical location of where he or she was searching.

For example, a user searching for information on the Empire State Building, Central Park, and Broadway tickets predicts a trip to New York. A hotel in the area could use that search history data to deliver a relevant and timely search related ad or message.

Analyze Consumer Behavior and Preference from Past Locations Visited

Location history of a consumer provides a lot of information specific to that person: where they like to shop, what they like to buy, how often they make the trip, and even how they get there. Obtaining this information gives great insight to marketers that enhances the ability to target consumers and deliver relevant, responsive location specific ads and information, even if the consumer is not currently in that area.

For example, a bagel shop might serve up a free coffee coupon to anyone who's visited a Starbucks location more than once within 10 blocks of its shop. The customers may be from anywhere in the city but their location history allows the bagel shop to target those who are likely to be in the area in the future.

Use Location-specific Landing Pages to Provide Relevant Content

It's important not only to target the right consumers, but to provide the most relevant information to them. If you find the right user who clicks on your ad, but the landing page for that ad isn't customized, that conversion could be lost. Offer different website landing pages for each targeted ad that match the reason that user was targeted.

Another way to get the right people to the right landing page is through geo-aware targeting. Your site or landing page can detect where the user is when they click on a banner or visit your website.

For example, if a user from a high income neighborhood visits a car dealer's site or clicks on a paid search display ad, that consumer may be directed to a landing page displaying a luxury vehicle, while consumers located in a lower income area may be targeted with a deal on an economy vehicle. The higher income consumers may be more interested in deals such as cash off or lower interest rates whereas those in lower income brackets may be more receptive to lower monthly payments.

Take Advantage of Geographic Specific Events

Lastly, geographic specific events, such as the weather or traditional local holiday celebrations, can be used to target consumers. Some events are known in advance, like St. Patrick's Day in Boston. Others are unexpected, like snow storms in Dallas.

Upon forecast of a blizzard, a hardware store may target consumers with content promoting snow shovels or snow blowers. The week before St. Patrick's Day, a clothing store may promote its green colored or festive attire. Either way, these events will spike demand for particular items and are a great opportunity to boost sales.

In summary, these are but a few of the examples of how geography plays such an important part in creating customized and targeted marketing campaigns. Consumers respond better to relevant marketing which means that ROI of targeted campaigns will increase. Mobile consumers make geography one of the best ways to target while technology and data make doing so a real advantage to those who use it. Sometimes it takes a little creativity, but it is worth the effort. Especially for the business of local.

Areas of Application

The main motive for targeting measures is the assumption that internet users consider web content tailored to match their online behavior to be more relevant and useful. One central factor in creating relevant content is the user's location. This data provides online retailers with references to the preferred language, sociocultural characteristics, and the legal framework for advertising in the respective region. But geo-targeting isn't just used for addressing target groups; as well as being used for adapting web content and preparing online advertisements, app developers also determine their customers' locations in order to implement cross-media strategies and link online channels with offline touch points. Geo-targeting techniques are used across various other industries, such as in the market economy, copyright protection, and securing online transactions.

- Multilingual web content: many websites are geared towards users all around the world and thus need to provide content in several different languages. Header information gives an indication of the preferred language, which is submitted automatically by the web browser following a server request. If a browser detects a page with several language options, the content management system plays the required version. So for example, while a website visitor based in the United States will see a page in English, a visitor looking at the same page in Brazil may see the same content in Portuguese. But not every website relies exclusively on the header information; geo-targeting also enables the browser to determine the website based on technical localization methods.

- Geo-targeting in e-commerce: web stores also use multilingual sites to address an international customer base. Geo-targeting can here offer the possibility to provide many different versions of the same store under one global URL. Depending on the target region, this may offer customers a different product range, region-specific currencies, prices, and terms of delivery as well as information on the nearest offline branch.

- Regional advertising: the advertising industry often makes use of geo-targeting in order to increase the relevance of online advertising for their chosen target group. Ad networks like Google Ad Words and Micro softs Bing Ads provide functions that enable businesses to focus on individual users and target groups using adverts for relevant products or settings. This can benefit businesses of all sizes; for example, small businesses can use this type of online advertising to attract new customers from their surrounding area. The aim of regional targeting is ultimately to minimize advertising expenditure and place the advertising material where it will be most effective.

- Location-based services: with the ever-increasing use of smartphones and tablets, geo-targeting also plays an important role in location-based services, (LBS). Many apps automatically record users' locations with their consent. This gives developers the opportunity to coordinate software functions with the user's location. It's possible then to create services that are available online within a company or in selected shops and restaurants. An example of LBS being put into practice is with coupon apps; if a customer enters a participating retailer, a discount notification will automatically pop up on the smartphone display.

- Market research: market research also relies heavily on geo data, particularly in the areas of defining target groups and narrowing down requests for products and services to specific geographical locations.

- Copyright protection: in copyright protection, a form of user localization known as geo blocking is used. This allows multimedia platforms such as YouTube to use geo-targeting options in order to limit content to specific countries or regions and in so doing, protecting the copyright. License fee-financed web content of public-law broadcasting corporations can only be called up in their countries of origin.

- Payment security: another area of application for user localization is online payment transactions. Transaction services use geo-targeting techniques to coordinate a user's location with their account data and thus detect any inconsistencies.

Practical Applications of Geotargeting

The central idea behind geo-targeting is that understanding consumer's real-time — or past —location helps marketers achieve the holy grail of delivering the "right message at the right time." In a simple example, an adult customer visiting car dealerships is likely interested in buying a car, and serving a local Honda ad to this customer more likely to be successful.

Serving a Honda ad to someone at (or near) a local car dealership is an example of geo-targeting on a hyper local level. But it applies on a larger scale, too.

In the print days, taking out an ad in the Detroit Free Press allowed businesses to know that primarily Detroit area residents who could actually visit the business would see the ad in question. Not so in the era of mobile ads which, if delivered indiscriminately and without location context, can be less successful because they aren't relevant or personal. Ad creative targeted at — and customized for — an Oklahoma consumer versus a New York City one is likely to be more effective in driving a physical sale.

In more sophisticated use cases, geo-targeting doesn't have to be solely based on a consumer's real-time location. Locations or businesses a customer has visited recently can be a great predictor of interests and intent, so adding targeting based on historical location as well can be key to delivering a captivating, relevant message.

Using Geotargeting

In an example of a particularly successful geo-targeted mobile ad campaign, Denny's partnered with location ad platform x-Ad in 2014.

x-Ad's first step was to expand its use of location data. "We wanted to move beyond just the where, and use location data to define who and what of audience targeting," Monica Ho, x-Ad's CMO, told Geo-Marketing in 2015. After all, while a consumer's proximity to a Denny's can be significant, targeting someone who has been to and enjoys their local Denny's will often prove more successful — whether or not they happen to be near the restaurant at the time they see the ad.

The tactic ended up paying dividends for Denny's. A targeted "Build Your Own Skillet" mobile ad campaign produced an 11.6 percent lift in in-store visits; a reprisal of the effort for "Build Your Own French Toast" delivered a 34 percent increase.

Geotargeting is a practice frequently deployed by such restaurants and brick-and-mortar businesses looking to drive local foot traffic, but it isn't exclusively the province of these verticals. Even sports teams have gotten in on the action, targeting fans that are at (or have been to) a particular stadium or event in order to drive ticket sales, app downloads, and more.

We can take advantage of geotargeting on the devices of fans in a particular stadium.

As brands continue to explore the possibilities of targeting users based on location, both real-time and historical, consumers will likely see more content that is truly relevant to their lives — and marketers can boost sales as a result.

References

- Kriging-interpolation-prediction: gisgeography.com, Retrieved 08 July 2018

- The-nonlinear-Markov-Chain-Geostatistics-228890991: researchgate.net, Retrieved 19 July 2018

- Geodemographic-segmentation: spatially.com, Retrieved 11 April 2018

- What-is-geodemographic-segmentation: marketing91.com, Retrieved 05 July 2018

- How-does-geo-targeting-work-40: geoedge.com, Retrieved 25 May 2018

- 10-practical-tips-using-geo-location-reach-target-audience-217301: searchengineland.com, Retrieved 15 May 2018

- What-is-geo-targeting, web-analytics, online-marketing: 1and1.ca, Retrieved 15 March 2018

- What-is-geo-targeting, geomarketing-101: geomarketing.com, Retrieved 20 April 2018

6

Spatial Databases

Any database that is meant for storing and querying data representing objects in a geometric space is called a spatial database. These databases consist of geometric objects like polygons, points and lines for spatial representation. This chapter provides comprehensive knowledge of the different spatial database and database systems. It also includes an extensive discussion on object-based spatial database, spatiotemporal database and location intelligence, among others.

A spatial database is a database that is enhanced to store and access spatial data or data that defines a geometric space. These data are often associated with geographic locations and features, or constructed features like cities. Data on spatial databases are stored as coordinates, points, lines, polygons and topology. Some spatial databases handle more complex data like three-dimensional objects, topological coverage and linear networks.

Common database systems use indexes for a faster and more efficient search and access of data. This index, however, is not fit for spatial queries. Instead, spatial databases use something like a unique index called a spatial index to speed up database performance. Spatial indexing is very much required because a system should be able to retrieve data from a large collection of objects without really searching the whole bunch. It should also support relationships between connecting objects from different classes in a better manner than just filtering.

Aside from the indexes, spatial databases also offer spatial data types in their data model and query language. These databases require special kinds of data types to provide a fundamental abstraction and model the structure of the geometric objects with their corresponding relationships and operations in the spatial environment. Without these kind of data types, the system would not be able to support the kind of modeling a spatial database offers.

Spatial Database System

A spatial database system may be defined as a database system that offers spatial data types in its data model and query language, and supports spatial data types in its implementation, providing at least spatial indexing and spatial join methods.

Spatial database systems offer the underlying database technology for geographic information systems and other applications.

In various fields there is a need to manage geometric, geographic, or spatial data, which means data related to space. The space of interest can be, for example, the two dimensional abstraction of parts of the surface of the earth or a 3d-space representing a digital terrain model. At least since the advent of relational database systems there have been attempts to manage such data in database systems.

Characteristic for the technology emerging to address these needs is the capability to deal with large collections of relatively simple geometric objects, for example, a set of 100 000 polygons. Several terms have been used for database systems offering such support like pictorial, image, geometric, geographic, or spatial database system. The terms "pictorial" and "image" database system arise from the fact that the data to be managed are often initially captured in the form of digital raster images (e.g. remote sensing by satellites, or computer tomography in medical applications).

The term "spatial database system" has become popular during the last few years, and is associated with a view of a database as containing sets of objects in space rather than images or pictures of a space. Indeed, the requirements and techniques for dealing with objects in space that have identity and well-defined extents, locations, and relationships are rather different from those for dealing with raster images.

A spatial database therefore has the following characteristics:

- A spatial database system is a database system.

- It offers spatial data types (SDTs) in its data model and query language.

- It supports spatial data types in its implementation, providing at least spatial indexing and efficient algorithms for spatial join.

Nobody cares about a special purpose system that is not able to handle all the standard data modeling and querying tasks. Hence a spatial database system is a full-fledged database system with additional capabilities for handling spatial data. Therefore spatial indexing is mandatory. It should also support connecting objects from different classes through some spatial relationship.

Spatial Database Design

A spatial database includes collections of information about the spatial location, relationship and shape of topological geographic features and the data in the form of attributes. The design of the spatial database is the formal process of analyzing facts about the real world into a structured model. Database design is characterized by the following phases: requirement analysis, logical design and physical design. In other words, you basically need a plan, a design layout and then the data to complete the process.

Having a solid well designed spatial database is the key to performing good spatial analysis. The database can be complex and designed with expensive sophisticated software or can be merely a simple well organized collection of data that can be utilized in a geographic form.

Three main categories of spatial modeling functions that can be applied to geographic features within a GIS are:

(1) Geometric models, such as calculating the Euclidean distance between features, generating buffers, calculating areas and perimeters and so on;

(2) Coincidence models, such as topological overlay;

(3) Adjacency models (path finding, redistricting, and allocation).

All three model categories support operations on spatial data such as points, lines, polygons, tins, and grids. Functions are organized in a sequence of steps to derive the desired information for analysis.

Almost all entities of geographic reality have at least a 3-dimensional spatial character, but not all dimensions may be needed. E.g. a highway pavement actually has a depth which might be important, but is not as important as the width, which is not as important as the length. Representation should be based on the types of manipulations that might be undertaken. Map-scale of the source document is important in constraining the level of detail represented in a database. E.g. on a 1:100,000 map individual houses or fields are not visible.

Steps in Database Design

- Conceptual
  - Software and hardware independent.
  - Describes and defines included entities.
  - Identifies how entities will be represented in the database.
  - I.e. selection of spatial objects - points, lines, areas, raster cells.
  - Requires decisions about how real-world dimensionality and relationships will be represented.
  - These can be based on the processing that will be done on these objects.
  - E.g. should a building be represented as an area or a point?
  - E.g. should highway segments be explicitly linked in the database?
- Logical
  - Software specific but hardware independent.
  - Sets out the logical structure of the database elements, determined by the data. Base management system used by the software.
- Physical
  - Both hardware and software specific.
  - Requires consideration of how files will be structured for access from the disk.

Characteristics of a Good Database Design

In order that the GIS database provides the best service it should be:

- Contemporaneous – the data should be updated regularly so as to yield information that pertains to the same time-frame for all its measured variables.
- Flexible and extensible so that additional datasets may be added as necessary for the intended applications:

- o The categories of information and subcategories within them should contain all of the data needed to analyze or model the behavior of the resource using conventional methods and models.

- Positionally accurate – if for example the boundary between the residential and agricultural land has changed, this may be incorporated with ease.

- Exactly compatible with other information that may be overlain with it.

- Internally accurate, portraying the nature of phenomena without error - requires clear definitions of phenomena that are included.

- Readily updated on a regular schedule.

- Accessible to whoever needs it.

Spatial Database Management

Many factors influence a successful Geographic Information System (GIS) implementation. None however are more fundamental than having the right management strategies and software to implement these. The spatial database is the foundation by which all data is uniformly created and converted. But maintaining the integrity and currency of the data is of fundamental importance. A classic mistake made by many organizations is thinking that a generic spatial database design will be sufficient for their needs. That is simply not the case. The spatial database is the end result of a series of processes that determine the specific functional requirements for the user and the key applications. Interoperability of data is also a critical area of concern in the development of spatial data information systems. As we move from newly created data to assimilation of all existing data, a properly designed spatial database is insurance for end user success. A good spatial database management software package should be able to:

- Scale and rotate coordinate values for "best fit" projection overlays and changes.

- Convert (interchange) between polygon and grid formats.

- Permit rapid updating, allowing data changes with relative ease.

- Allow for multiple users and multiple interactions between compatible data bases.

- Retrieve, transform, and combine data elements efficiently.

- Search, identify, and route a variety of different data items and score these values with assigned weighted values, to facilitate proximity and routing analysis.

- Perform statistical analysis, such as multivariate regression, correlations etc.

- Overlay one file variable onto another, i.e., map super positioning.

- Measure area, distance, and association between points and fields.

- Model and simulate, and formulate predictive scenarios, in a fashion that allows for direct interactions between the user group and the computer program.

Object-based Spatial Database

The choice of an appropriate representation for the structure of a problem is perhaps the most important component of its solution. For database design, the means of representation is provided by the data model. A data model provides a tool for specifying the structural and behavioural properties of a database and ideally should provide a language which allows the user and database designer to express their requirements in ways that they find appropriate, while being capable of transformation to structures suitable for implementation in a database management system. Data modelling is among the first stages of database design. The purpose of data modelling is to bring about the design of a database which performs efficiently; contains correct information (and which makes the entry of incorrect data as difficult as possible); whose logical structure is natural enough to be understood by users; and is as easy as possible to maintain and extend. Of course, different problems require different means of representation and a large number of data models is described in the database literature. Some are close to implementation structures, for example the relational model. Others as yet have no directly corresponding implementation. This is the case for those which support a wide variety of modelling constructs as well as a high level of abstraction. Such models allow representations which are closer to the original problem as framed by the user. An example is the IFO (Is-a relationships, Functional relationships, complex Objects) model.

Semantic Data Models

The relational model provides the database designer with a modelling tool which is independent of the details of physical implementation. However, the relational model is limited with respect to semantic content (i.e., expressive power) and there are many design problems which are not naturally expressible in terms of relations. Spatial systems are a case where the limitations become clear. To illustrate this point, consider the relational model of a polygon as originally given by van Roessel based upon the definitions proposed by the National Committee for Digital Cartographic Data Standards.

- Polygon (Polygon, ID, Ring ID, Ring Seq);

- Ring (Ring ID, Chain ID, Chain Seq);

- Chain1 (Chain ID, Point ID, Point Seq);

- Chain2 (Chain ID, Start Node, End Node, Left Pol, Right Pol);

- Node (Node ID, Point ID);

- Point (Point ID, X Coord, Y Coord).

This model of a polygon as a set of relations, though complete, is low-level and some way from one which represents a user's normal view of such an object. Semantic data models aim to provide more facilities for the representation of the users' view of systems than the relational model, as well as to de-couple these representations from the physical implementation of the databases. Fundamental work in this area was undertaken by Chen, who proposed a semantic data model and a diagrammatic technique known as the entity-relationship (E-R) model and diagram respectively.

Entity-relationship Modelling

The entity-relationship model utilizes the concepts of entity, attribute and relationship. A distinction is made between a type and an occurrence of a type. An entity is an item about which the database is to record information. Such an item should be uniquely identifiable. For example, a particular point could be uniquely identified by its coordinates or a census tract by its census code. An entity type is an abstraction representing a class of entities of the same kind. For example, POINT and CITY are entity types. Occurrences of those types are particular points and cities, e.g., a point with coordinates and a city named Oxford. An attribute is an element of data associated with an entity. A city has a population, thus the entity type CITY has attribute type POPULATION.

Figure: Entities and relationship

A particular city has a particular population. Such a population is an example of an attributed occurrence. To avoid a cumbersome presentation, we will omit the terms 'type' and 'occurrence' when no ambiguity is involved. The attribute(s) which identify an occurrence of an entity uniquely are termed identifiers or keys.

A relationship is an association between entities. For example, LIVES_IN is relationship between entities PERSON and CITY. Again, we may distinguish between types and occurrences of relationships. Relationships may have attributes, for example, the relationship LIVES_IN might have the attribute DURATION which gives the length of time that a person has lived in a city.

Chen proposed a diagrammatic means of representing this model. The diagrammatic form of the above example is shown in figure Rectangles depict entities and rhombi depict relationships. M and N indicate that the relationship is many-to-many, that is each person may live in more than one city and each city may have more than one person living in it.

Many systems may be modelled using entities, attributes and relationships, including systems with a dominating spatial component. Calkins and Marble (1987) apply the method to the design of a cartographic database. They describe the strengths of the method as being flexibility, control of database integrity and generality.

An important feature of E-R modelling is the natural and well-understood method of the transformation from the E-R model to the rational model.

Extensions to the Entity-relationship Model

The entity-relationship approach is at present recognized as the prime tool for data modelling. However, experience has shown that for many systems the initial set of modelling constructs (entity, attribute and relationship) is inadequate. For example, view integration (the process by which several local views are merged into a single integrated model of the database) is recognized by many workers as of great importance for GIS design. View integration is greatly facilitated by the introduction of abstraction concepts additional to the original E-R model. In the mid-1970s, Smith and Smith proposed the introduction of two abstraction constructs, generalization and aggregation, into the data modelling tool-kit. These constructs are provided by almost all contemporary semantic data models. We proceed to describe them in detail. distinguishing between generalization and specialization, considering a further construct, association or grouping, and then discussing their relevance to spatial database design. A similar approach will be found in Egenhofer and Frank (1989).

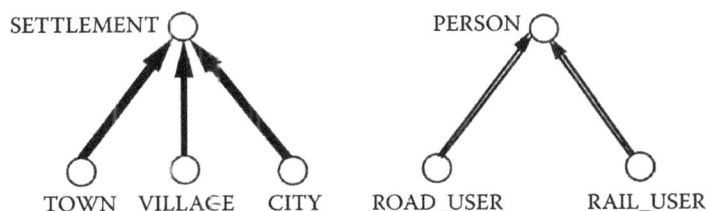

Figure: SETTLEMENT is a generalization of TOWN, VILLAGE and CITY. ROAD_ and RAIL_USER are specialization of PERSON

Generalization

Generalization is the construct which enables groups of entities of similar types to be considered as a single higher-order type. For example, entities of types VILLAGE, TOWN, and CITY may be merged and considered as entities of the single type SETTLEMENT. SETTLEMENT is said to be a generalization of VILLAGE, TOWN, and CITY. The diagrammatic representation of this situation is shown in figure above Formally, the generic higher-order type is the set-theoretic union of objects in the lower-order types. An object may be thought of, at this intermediate stage between classical and object-oriented data modelling, as an entity along with its attributes.

Specialization

Specialization is the construct which enables the modeller to define possible roles for members of a given type. For example, entities of type PERSON might be considered occurrences of type ROAD_USER or RAIL_USER, depending upon the context in which we see them. The diagrammatic representation of this situation is shown in figure above formally, the specialized type is made up of a subset of occurrences of the higher-order type.

It should be noted that, although generalization and specialization are in a sense inverse to one another, there are distinctions. A generic type inherits its structure from its lower-order types (and possibly adds some of its own). In the case of specialization, the specific types inherit structure from the higher-order type (and possibly add some of their own). In the case of both generalization and specialization, we say that the lower-order type is a subtype of the higher-order type.

Aggregation

Aggregation is the construct which enables types to be amalgamated into a higher order type, the attributes of whose objects are a combination of the attributes of the objects of the constituent types. Formally, the objects which are occurrences of the aggregate type are tuples, the components of which are the objects of the constituent types. In short, aggregation corresponds to the mathematical operation of Cartesian product. An example is the type POINT, which is the aggregate of type POINT_ID with two integer types named X_COORD and Y_COORD, thus representing a point as having two spatial coordinates. This relationship is represented diagrammatically in figure above.

ROAD is an ordered association of INTERSECTION.

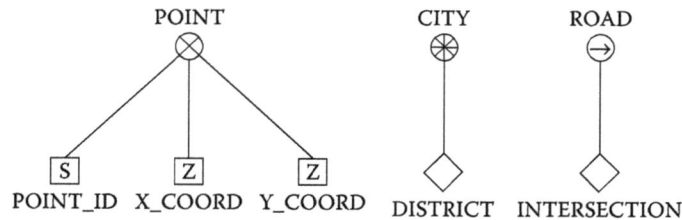

Figure: POINT is an aggregation of identifier and coordinates. CITY is an association of DISTRICT.

Association

Association or grouping is the construct which enables a set of objects of the same type to form an object of higher level type. It is often stipulated that the sets are finite. The corresponding set-theoretic construct is the power-set operator. An example is the view of a city as, amongst other things, a collection of districts. CITY is an association of DISTRICT. This is shown diagrammatically in figure above.

Ordered Association

Sometimes, it is important to take into account the ordering of a collection of objects. For example, the ordering of intersections making up a road may be critical. Ordered sets are called lists and we allow a higher-order type which is a collection of lists of the lower-order type. In our example we say that ROAD is an ordered association of INTERSECTION. This is shown diagrammatically in figure above.

Object-oriented Data Modelling

In describing the above abstraction constructs, we have gradually moved towards an object-oriented view. In the basic E-R model, the entities are conceived as having attributes, occurrences of which are drawn from atomic domains. That is, the underlying domains are of basic and indecomposable types such as INTEGER, REAL and STRING. As we bring in the abstractions above, we add a further dimension to the structure of the underlying domains, which no longer need be atomic.

In object-oriented data modelling, all conceptual entities are modelled as objects. An abstraction representing a collection of objects with properties in common is called an object type. Objects of the same type share common functions. The objects associated with an object type are called occurrences. INTEGER and STRING are object types, as is a complex assembly such as a CITY. Indecomposable object types are called primitive. Decomposable objects are called composite or complex objects.

A composite object, therefore, is an object with a hierarchy of component objects. We have seen how complex types may be formed from primitive types using generalization, specialization, aggregation and grouping. These are the primary Object-Oriented Data Modelling for Spatial Databases 125 object-type operations in object-oriented data modelling. Other operations have been introduced and can be found in the literature.

Object-oriented data models support the description of both the structural and behavioural properties of a database. Structural properties concern the static organizational nature of the database. Behavioural properties are dynamic and concern nature of possible allowable changes to the information in the database. This paper concentrates on the structural description.

The object-oriented approach to data modelling has proved to be especially fruitful in application areas which are not of the standard corporate database type. Complex molecular and engineering part-assembly databases are examples of systems which have been successfully modelled using these techniques. What such applications have in common is a richly-structured semantic domain, often with a hierarchical emphasis, and associated with multimedia database (e.g., text, numeric, graphical, audio). Since a GIS also shows these characteristics, it seems then that a GIS is an ideal application for object-oriented modelling.

In order to show more clearly how this methodology may be applied, we will concentrate on the specific recent data model IFO (Abiteboul and Hull 1987) which contains the above object-oriented constructs. It is concerned almost wholly with structural properties of a database.

IFO

The IFO model was introduced by Abiteboul and Hull (1984). A more condensed account of the model is given by Abiteboul and Hull (1987). IFO incorporates all the constructs so far introduced in this paper with the exception of 'relationship'.

Object Types

IFO is truly object-oriented in that all its component types may be composite. Atomic types are of three kinds; printable, abstract and free. A printable type corresponds to objects which may be represented directly as input and output. Examples of printable types are INTEGER, STRING, REAL and PIXEL. An abstract type (shown in IFO by a diamond) corresponds to physical or conceptual objects which are not printable. PERSON is an example of an abstract type. Free types (shown in IFO by circles) serve as links in generalization and specialization relationships. Representations of examples of atomic types are given in figure S and Z indicate STRING and INTEGER types respectively. Non-atomic types are constructed from atomic types using aggregation and grouping as already discussed. For diagrammatic clarity, it is sometimes convenient to treat complex types as atomic. For example, figure below shows POINT is an aggregate of atomic types but later diagrams treat POINT as abstract atomic.

POINT_ID   X_COORD   PERSON   SETTLEMENT

Figure: Atomic types.

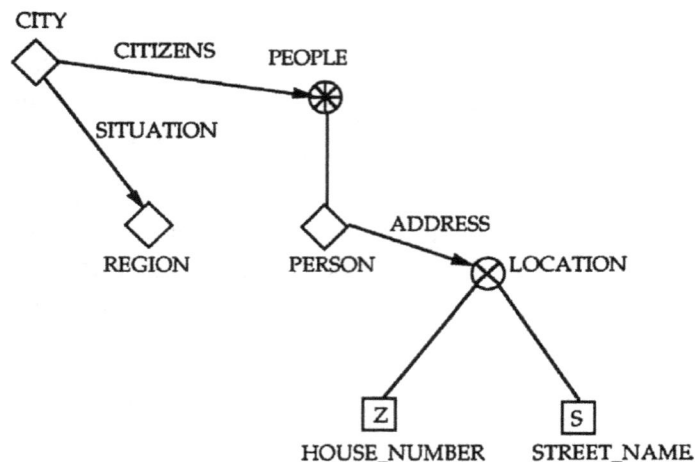

Figure: The CITY fragment if IFO

Functional Relationships between Objects

So far the ways in which complex objects may be constructed from atoms have been described. We now discuss how types may be related. IFO provides formalism for representing functional relationships between types. The means by which functional relationships are represented is the fragment. Informally, a fragment is a part of the IFO model, containing types and functions (but no generalization or specialization relationships), subject to certain constraints. We illustrate with an example, shown in figure above This fragment shows functional relationships SITUATION and CITIZENS between object types CITY, REGION and PEOPLE. The structure of the knowledge being modelled here is that cities are situated in regions and are occupied by people, each of whom may have for an address a location which is an aggregation of a house number and street name. The function CITIZENS has the dependent function ADDRESS. Intuitively, this models the case where a person may live in more than one city and so have different addresses in different cities.

The E-R model allows the possibility of many-valued relationships between types and so appears to be more general than a functional model. However, the grouping operator can be used to provide the facility of representing many-valued functions. For example, the relationship shown in figure, where a person may live in several cities and a city comprises many people, is represented functionally in IFO as shown in figure, where the image of a city under the function CITIZENS is a set of persons, since it is an object of type PEOPLE, which is an association of PERSON. Formally, a fragment F is a rooted directed tree of types. The root of a fragment is called a primary vertex.

Schemas

In IFO, fragments form the building blocks of schemas. A schema is the largest IFO unit and is a forest of fragments, possibly connected together at their primary vertices.

Object-Oriented data modelling for spatial databases by generalization and specialization edges. Thus the schema allows the representation of all the components of the IFO model.

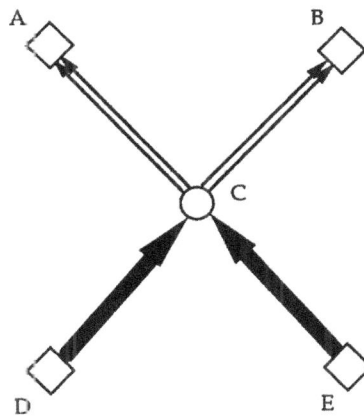

Figure: An inconsistent structuring of objects

IFO places some constraints on the way that schemata may be constructed. These concern the directions in which objects are structured from other objects. For example, aggregated and grouped objects are constructed from their constituent elements. The non commutativity of this structuring relationship leads to the distinction made between generalization and specialization. A generalized type is structured from its sub-types. A specialized type is structured from its super-types. The schema must have its object sources and sinks arranged in a consistent fashion. For example, the configuration shown in figure is not permissible as it results in inconsistent structuring of objects. Objects of type C result from specializing objects of types A and B and generalizing objects of types D and E. There is no guarantee that this can be done consistently and that clashes will not result. A further point here is that, even neglecting D and E, C is the result of specializing from two possibly quite distinct types A and B. Again, there is no guarantee that this can be done consistently and we would require that types A and B 'arise' from a common type. If all the arrows in figure were reversed, we would again have an impermissible schema. In this case, C is a source for objects of types A, B, D and E, but is not itself defined in terms of any other type. It could be considered a 'black hole' of the system. Abiteboul and Hull define the permissible configurations by stating five rules which they must satisfy. They also state a theorem showing that any schema satisfying their rules leads to a consistent structuring of objects at each vertex of the schema.

Fundamental Spatial Objects of IFO

The first application of IFO that we present is its use to represent the three fundamental spatial object types; point, line, and polygon. These representations are based upon the definitions proposed by the National Committee for Digital Cartographic Data Standards, which are summarized in van Roessel as follows:

- A point is a zero-dimensional spatial object with coordinates and a unique identifier within the map.

- A line is a sequence of ordered points, where the beginning of the line may have a special start node and the end a special end node.

- A chain is a line which is a part of one or more polygons and therefore also has a left and right polygon identifier in addition to the start and end node.

- A node is a junction or endpoint of one or more lines or chains.

- A ring consists of one or more chains.

- A polygon consists of one outer and zero or more inner rings.

It is shown earlier how it is possible to represent these spatial elements directly using the relational model. An IFO representation of POINT is given in figure and that for NODE, LINE and POLYGON in figures, respectively. Shows a node as a special kind of point with its own node identifier as well as its identifier as a point. In figure below a line is modelled as an ordered association of points with identifier and begin and end nodes. A polygon, in figure below, is an ordered association of rings, which in turn are ordered associations of chains. Polygons, rings and chains have identifiers. A chain is a special type of line with corresponding left and right polygons.

The aim is a presentation which accords with users' own views of an object and decouples the representation from the implementation.

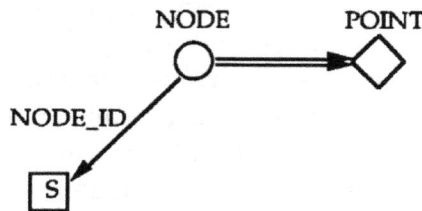

Figure: NODE modelled in IFO

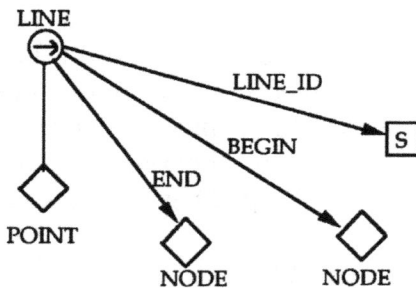

Figure: LINE modelled in IFO

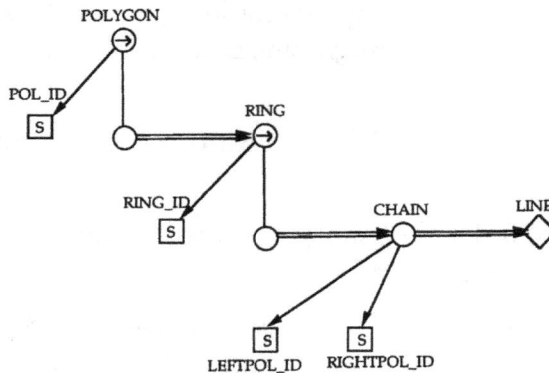

Figure: POLYGON modelled in IFO

Post-code System

The United Kingdom is divided into 120 post-code areas, of which Leicester is one, designated LE. The post-code area, LE, is divided into 17 district post-codes (LE1 to LE17). These in turn are divided into 86 sector post-codes. Sectors are divided into the smallest of the post-code areal units: the unit post-code (UPC) e.g., LE1 7RH, which is a unique identifier for all the points of delivery (addresses) on one postman's 'walk'. The post-code address file (PAF) provides the UPC for all addresses in the county. It also provides a coded form of the address: a 4-digit PREMCODE, which consists of the first four characters of information of an address, e.g., 3 Main Street has PREMCODE 3MAI. There are provisions in the PREMCODE for removing ambiguities, and so this, together with the UPC, can uniquely identify every address. It can be seen that these units nest neatly into each other in a clearly defined, and clearly identifiable, hierarchy.

Figure: Map of Great Britain showing Leicestershire. Shaded areas show land above 200 m

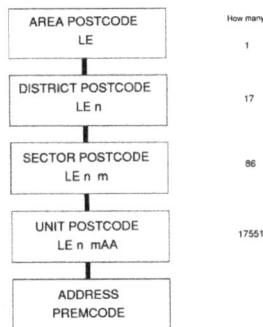

Figure: The post-code units for Leicestershire

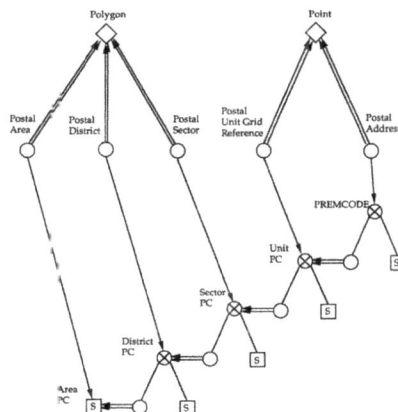

Figure: Relationship between postal units in IFO

The Central Postcode Directory (CPD) provides the Ordnance Survey Grid reference for the SW corner of a 100m grid square for the first address in each UPC.

Figure displays the postal units in IFO. POLYGON and POINT types, defined earlier, are taken as given abstract atomic types. It can be seen that each postal unit type (for, example, POSTAL DISTRICT) is a specialization of an abstract spatial type (for example, POLYGON), and is also the domain of a function whose co-domain is a composite string type which constitutes an identifier (for example, DISTRICT POSTCODE). This representation makes a clear and useful distinction between the spatial and non-spatial aspects of the postal units.

Spatiotemporal Database

A spatiotemporal database is a new type of database system that manages spatiotemporal objects and supports corresponding query functionalities. A spatiotemporal object is a kind of object that dynamically updates spatial locations and/or extents along with time. A typical example of a spatiotemporal object is a moving object (e.g., a car, a flight or a pedestrian) whose location continuously changes.

Spatiotemporal databases have many important applications such as Geographic Information Systems, Location-aware Systems, Traffic Monitoring Systems, and Environmental Information Systems. Due to their importance, spatiotemporal database systems are very actively researched in the database domain.

Contentions Relating to Spatiotemporal Databases

Spatiotemporal databases are not (just) about moving objects. There has been significant recent activity on moving objects (e.g.,); such work seems likely to yield results that are of value to real applications. However, it is by no means obvious that moving object proposals will address all the requirements of applications modelling changes to objects that are not directly associated with movement (e.g., boundaries, rivers).

Spatial and a spatial data change in similar ways. This contention essentially proposes that a single temporal model should be usable with both spatial and a spatial data.

There is more to query processing than indexes. There has been much more research carried out on spatiotemporal indexes than on other features of query processors – query algebras, optimisers, join algorithms, etc. even though these components are crucial to the development of complete spatiotemporal database systems.

Rampant featurism has been a significant impediment to real progress. Proposals for spatial and temporal database models and languages often have very large numbers of features, which together would be extremely difficult to implement. More is not always better, and proposals that are difficult to implement may impede the development of leaner, more practical approaches.

You can't please all of the people any of the time so it is enough to start by pleasing someone. This follows on from the previous contention. In the absence of practical spatiotemporal database systems, users of spatial and temporal data are making do with much more generic facilities. A significant number of tasks can probably be given effective support by reasonably lean, but nevertheless fully developed, spatiotemporal systems.

It is easier to propose a spatiotemporal model than to implement it. We assume that this is the reason why there are so many more proposals than prototypes. Researchers should be slow to propose models that they are not prepared to prototype (and thereby to evaluate in practice).

Spatiotemporal databases are not descended from geographical information systems. The right place to start building a spatiotemporal database system is a spatial database system, not a GIS. Adding facilities that are characteristic of temporal databases (e.g., declarative queries) to a GIS is likely to be a rocky road to a cumbersome proposal.

The above contentions motivate a view that research in spatiotemporal databases could benefit from a period of consolidation, in which the focus is on the development of comprehensive working prototypes. Such an emphasis can be used to identify areas that present open research challenges that must be addressed before effective spatiotemporal database systems can be developed. Such an emphasis also focuses the mind in terms of the most important kinds of functionality – it is unlikely that scalable working prototypes can be developed in the short to medium term that support all of vector and raster data, valid and transaction time, spatial and temporal indeterminacy, moving objects and constraint programming. Thus identifying functionalities that together are useful for significant categories of application and focusing on them, should allow individual research activities to identify important challenges that are grounded on well-defined application needs.

Components

Architectural View of DBMSs

We propose an architectural framework in which DBMSs are viewed as comprising programming language interfaces, a query processing component, and a services manager (which, we assume, is responsible for the management of storage structures, security, concurrency, transactions, recovery, etc.).

The programming language interfaces allow application programs to access and manipulate data using the abstractions of the data model implemented by the DBMS.

The query processor is assumed to comprise a compiler (from a surface syntax to a logical algebra), a logical optimiser (which uses equivalences to rewrite logical algebraic expressions into forms that are, heuristically, more efficient to evaluate), a physical optimiser (which, in the light of the available access methods and indices, uses statistics about the data to convert logical algebraic operators into physical algebraic ones, thereby generating a query plan), and an evaluator (which traverses the query plan making calls to algorithms and access methods to compute the answer).

The services manager is assumed to be visible via an application program interface (API) which serves data and metadata to application programs and to the query processor whilst enforcing, e.g., concurrency, transaction, recovery and security controls.

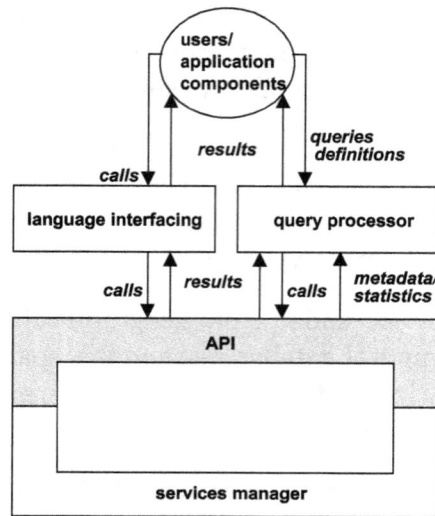

Figure: A Component-Based DBMS Architecture

Spatial and Temporal Extensions

We now consider the problem of specifying, designing and implementing spatial and temporal DBMS by extending architecture such as that depicted in figures. The targeted DBMS is expected to support DBMS functionality over spatial and temporal data orthogonally and synergistically.

By orthogonality we mean that users can model the spatial features of types and properties, or the temporal aspects of modelled types and properties, or both, or neither.

By synergy we mean that if a user makes use of either only spatial or only temporal or both spatial and temporal facilities, the system responds positively (e.g., by making available specific syntax, by providing additional behaviour, by seeking optimisation opportunities, etc.).

The most essential decision in this approach to extensions is, of course, how the a spatial, snapshot data model is to be impacted.

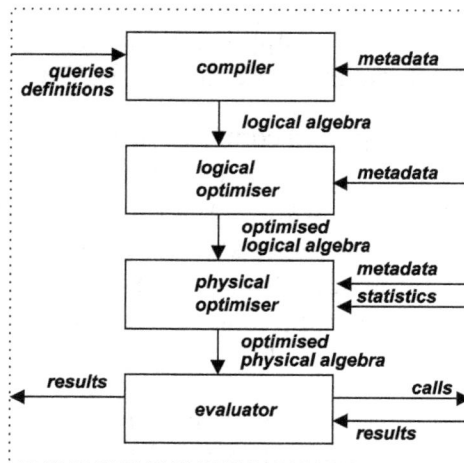

Figure: The Query Processor

We note that, at the level of the data model, extending an spatial database (temporal or not) to support spatial applications only requires extending the set of supported types (to include, say, points, lines and regions). By contrast, extending a snapshot database (spatial or not) to support temporal applications requires not only extending the set of supported types (to include, say, instants and intervals), but also making provisions for all (or at least many) data model types to be associated with a history. In a sense, spatial extensions need induce no changes to the pre-extension types, whereas temporal extensions imply that the way the DBMS used to handle types is different from the way that it is expected to handle them now.

Figuratively speaking, spatial extensions can be layered over pre-extension types, but temporal extensions, at the very least, wrap around all supported types. The same applies, of course, to index and access method infrastructures. This view of the world is illustrated in figure above.

Implications of Spatial Extensions

Extensions to support spatial applications are backward compatible with the set of types provided by the systems being extended in the sense that their specification, design and implementation depends on that of the existing types (i.e., integers, strings, etc.) but does not alter them at any level of abstraction (i.e., conceptually, logically or physically).

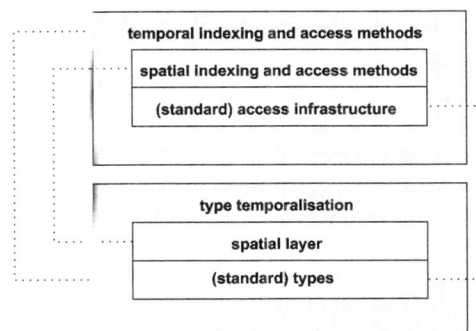

Figure: The storage manager in a spatiotemporal services manager component

Given a set of spatial types, their formal characterisation (e.g., as carriers of a spatial algebra) defines well formed-ness conditions for the insertion and deletion of values in database states, relationships between types (which can inform the design of spatial indexing mechanisms and optimisation strategies), and, most visibly, the operations that can be used to access and mutate values, as well as to test for conditions and to map values of one type into values of some other type.

This characterisation of spatial types as new primitive types gives rise on the one hand to a concrete syntax (for data definition and manipulation) and to a logical algebra. The spatial logical algebra is integrated with the existing logical algebra as yet another primitive algebra (on a par, therefore, with the integer algebra, the string algebra, etc.) among those that the services manager supports. Thus, just as addition on integers can be operator nodes in a query tree, so can, say, union of regions. Likewise, just as a less-than predicate on integer values can appear as a selection condition, so can, say, an inside predicate on, e.g., a lines and a regions value.

For a services manager to support a spatial algebra it must implement storage structures for spatial values, algorithms (denoted by physical algebraic operators) for the spatial operations, and auxiliary structures (such as spatial indices). Therefore, new services are added or extended, but no existing services need to be significantly disrupted. As figure indicates, spatial extensions can, roughly speaking, be layered over existing functionality.

With respect to figure above, each and every module of the query processor is affected, but only by minimally disruptive extensions. Thus the compiler will recognise more literals, more function identifiers, and there will be more typing restrictions to verify and enforce. The logical and physical optimisers, as well as the evaluator, will range over an enlarged set of operators and will require new heuristics, and new cost modelling parameters in order to respond to the greater number of opportunities for rewriting and algorithm selection that arise.

Finally, with respect to figure, the language interfaces will be enriched with new signatures corresponding closely to the spatial algebraic specification adopted.

Implications of Temporal Extensions

Over and above analogous implications in each and all of the dimensions discussed above for the spatial case, temporal extensions also imply changes over the services previously provided for pre-extension types.

As in the spatial case, a set of temporal types (e.g., instants and intervals) must be formally characterised (e.g., as carriers of a temporal algebra) and, as described above, query processor components as well as language interfaces will be affected analogously.

Note that this only suffices for an application type (say, employee) to be associated with a property (say, employment-period, describing the (possibly closed) interval of time in which the employee has been with the employing organisation).

However, temporal extensions imply more functionality. In particular, the previous paragraph still characterises a snapshot database in that the only change to the snapshot model is an increase in the number of primitive types that are supported. In contrast to this, of course, a temporal DBMS keeps (if required) historical information, and thus information about the past is maintained by the system, and queries can take such information into account.

To support this richer view of how the objects modelled in the database are changing with time, temporal extensions impact all pre-extension types (including spatial types if they happen to be supported). For example, salary (of type, say, integer) may be temporalised (by which we mean that not just current but previous values too are maintained in a history). Similar considerations apply to values of any other type (e.g., address of type string, border of type region, etc.).

The type temporalisation referred to at the bottom of figure can conceptually be understood as the implicit availability of a new collection type for modelling histories, such that each element in a history (of an application type extent, property or relationship) pairs a value (of the extent, property or relationship) being modelled with a temporal value. In addition, histories come with an interface consisting of access or, filtering and conversion (but not necessarily constructor, mutator or destroyer) operations. The operations on histories in turn give rise to extensions to the logical and physical algebras of the query processor illustrated in figure.

Temporal extensions thus have significant consequences for the architecture of a database, in that the management of historical information on all types in the model implies significant revisions to the services manager and to the query processor.

Implications of Spatial Plus Temporal Extensions

Given the implications of spatial and of temporal extensions, which were considered in isolation above, what, if any, are the additional implications of having spatial types supported within a temporal DBMS? Strictly, the provision of a collection of spatial types within a temporal database should yield a coherent spatio-temporal database.

Assuming, for the meantime that a spatio-temporal database needs no operations over and above those associated with the spatial and temporal types, there may still be additional implications for the DBMS if good performance is to be obtained. For example, if a spatio-temporal index is to be used (e.g.,) to improve the performance of queries involving space and time, then this implies an extension to (at least) the physical algebra and optimiser, as well as the obvious introduction of the index to the services manager.

Furthermore, it is likely that, with a view to minimising the use of disk space for temporally evolving spatial data, specialised storage structures can be expected to be of benefit. For example, a storage manager could choose to record the modifications that have been made to a spatial value, rather than copying a large spatial value every time a small change is made.

Returning to the question of language extensions, it is possible, perhaps likely, that spatio-temporal databases would benefit from the provision of some language constructs in addition to those provided with independently conceived spatial and temporal extensions. These are not likely to be numerous, but arguments have been advanced to the effect that additional language constructs are useful in certain cases. Such constructs clearly require support through surface syntax, algebras and evaluation algorithms.

Location Intelligence

Location intelligence is an often heard term, but what it means and how it can benefit businesses, institutions and individuals may not be immediately obvious. Location intelligence is more than analysis of geospatial information or geographic information systems alone, it is the capability to visualize spatial data to identify and analyze relationships. Evolving from GIS, location intelligence provides analytic and operational solutions across organizations.

How does all this data help people, and what about the customer or client? Organizations have discovered that data can be one of the best ways to get insights about customers and how to serve them better, increasing brand loyalty and improving customer relationship management. Linking customer addresses to a geographic area and then running these against internal company data and external demographics such as census data and income data, or other open data can provide unprecedented levels of detail. Who people are, what they do, and how, when, they consume is tied to the where, in essential ways. What is their neighborhood, commute, and workplace?

These locations and their spatial relationships lead to a more in-depth understanding of behavior and influences. Since a high percentage of data already has geographical information attached to it, insights about these relationships are readily available. Location intelligence now allows for incorporating external data from a variety of sources that can be combined and updated dynamically in the cloud. Companies can update the accessibility of their brand locations, marketing and potential new sites accordingly.

More than ever location intelligence has made it easy to map excel data, create data dashboards, and derive deep insights from location data. Discover how real companies across a range of industries and categories: finance, real estate, economic development and operational logistics are making use of location intelligence technology to gain a competitive edge in the marketplace.

Customer Insights

Location intelligence (LI) enriches traditional data with demographic or lifestyle data by adding spatial data metrics such as orders per location, e-mail clicks per region, etc. to use in sales forecasting. Location data can also be used in customer profiling, segmentation and customer habits. Data driven maps can also be used to see how the demographics and buying habits of customers located around the stores have changed over time.

Marketing and Advertising

Since a high percentage of data already has geographical information attached to it—thanks to smartphones and tablets—customers are no longer identified only by a post code and a home telephone number. Effective marketing messages can now be tailored based on customers' historical location data as well as real-time location in order to provide personalized solutions.

Choosing a Location for your Business

Location intelligence tools can help retailers to optimize the siting of new stores, dealerships, branch offices, factories etc. Moreover, existing stores can be prioritized and geographic gaps can be filled, based on local demographic, economic data and foot-traffic in the given location.

When Peugeot-Citroen (PC) needed to optimize their catchment areas, they decided not to use their existing system which relied on sending their fleet of vehicles in order to determine the driving time to each catchment location. Instead, they used LI tools to optimize the driving times which resulted in higher accuracy and lower cost.

Supply-chain and Risk Management

Global companies, in particular, need LI systems to mitigate risks such as natural disasters, political instabilities etc. which might negatively affect their production, supply and delivery chains. LI systems enable to visualize supply-chain networks and develop more accurate risk management plans. In the financial services sector, insurance companies are using LI systems to simulate natural disaster situations and estimate the potential pay out.

Mobile Asset Tracking

Large companies with extensive number of mobile assets and personnel use LI to track their equipment and workers in the field. Tablets, smartphones and smart sensors enable businesses to track vehicles and workers in real time. In addition to real-time tracking, Location Intelligence tools can notify supervisors if the assets leave or enter a particular location in addition to gathering system status data which can be used for proactive equipment maintenance or disaster mitigation.

The use cases for location intelligence tools are essentially endless, and businesses discover new ones in as the needs evolve. Taking advantage of LI should be one of the key priorities for decision makers seeking quality insights in an increasingly competitive global market. Done effectively, LI can help lower operating costs, maximize customer satisfaction and mitigate business risk.

Permissions

Index